S0-FFQ-950

Springer Texts in Political Science and International Relations

Springer Texts in Political Science and International Relations delivers high-quality instructional content for undergraduates and graduates in all areas of Political Science and International Relations. The series is comprised of self-contained books with a broad and comprehensive coverage that are suitable for class as well as for individual self-study. All texts are authored by established experts in their fields and offer a solid methodological background, often accompanied by problems and exercises.

Daniel Stockemer · Jean-Nicolas Bordeleau

Quantitative Methods for the Social Sciences

A Practical Introduction with Examples in R

Second Edition

Springer

/aniel Stockemer
School of Political Studies
University of Ottawa
Ottawa, ON, Canada

Jean-Nicolas Bordeleau
School of Political Studies
University of Ottawa
Ottawa, ON, Canada

ISSN 2730-955X ISSN 2730-9568 (electronic)
Springer Texts in Political Science and International Relations
ISBN 978-3-031-34582-1 ISBN 978-3-031-34583-8 (eBook)
https://doi.org/10.1007/978-3-031-34583-8

1st edition: © Springer International Publishing AG 2019
2nd edition: © The Editor(s) (if applicable) and The Author(s), under exclusive license to Springer Nature Switzerland AG 2023

This work is subject to copyright. All rights are solely and exclusively licensed by the Publisher, whether the whole or part of the material is concerned, specifically the rights of translation, reprinting, reuse of illustrations, recitation, broadcasting, reproduction on microfilms or in any other physical way, and transmission or information storage and retrieval, electronic adaptation, computer software, or by similar or dissimilar methodology now known or hereafter developed.
The use of general descriptive names, registered names, trademarks, service marks, etc. in this publication does not imply, even in the absence of a specific statement, that such names are exempt from the relevant protective laws and regulations and therefore free for general use.
The publisher, the authors, and the editors are safe to assume that the advice and information in this book are believed to be true and accurate at the date of publication. Neither the publisher nor the authors or the editors give a warranty, expressed or implied, with respect to the material contained herein or for any errors or omissions that may have been made. The publisher remains neutral with regard to jurisdictional claims in published maps and institutional affiliations.

This Springer imprint is published by the registered company Springer Nature Switzerland AG
The registered company address is: Gewerbestrasse 11, 6330 Cham, Switzerland

Preface

Welcome to the second edition of the book *Quantitative Methods for the Social Sciences: A Practical Introduction with Examples in SPSS and Stata.* This second edition complements its predecessor by introducing readers to a third statistical software. As its title makes clear, the first edition focused on two widely used statistical programs: the Statistical Package for the Social Sciences (SPSS) and Stata.

In the present edition, we seek to introduce quantitative methods for the social sciences using a third data science program: the R Project for Statistical Computing. R is one of the fastest-growing programming language in the data sciences. More and more, academics and industry researchers rely on R for their data analytical needs. There are many reasons to use R, chief of which are its accessibility, open-source system, and its package-based operations. Unlike many other programs in quantitative methods, R is free and easily accessible to everyone, two features that make it an attractive alternative to other expensive programs. Further, R operates based on open-source packages which are constructed and continuously updated by a large community of statisticians and users. In other words, the R programming language serves as the foundation for the development of various packages which can be used to organize and analyze data. Accordingly, R is in constant evolution with new packages being built continuously as well as important updates being made available for existing packages.

We hope that this book can help undergraduate and graduate students build a solid foundation in quantitative research methods in the social sciences. Moreover, we believe that this new edition can be a guide for quantitative researchers in the social sciences, specifically those who seek to work with R or transition from other programs to R. We also welcome any feedback regarding this second edition and wish to promote our engagement with readers.

Lastly, we wish to thank our colleagues Philippe Chassé, Juliette Leblanc, and Matthew Taylor at the Université de Montréal for their help in reviewing earlier versions of this textbook.

Happy learning,

Ottawa, Canada

Dr. Daniel Stockemer
Mr. Jean-Nicolas Bordeleau

Contents

About the Authors

Daniel Stockemer is Full Professor in the School of Political Studies at the University of Ottawa and since May 2021 chair holder of the Konrad Adenauer Research Chair in Empirical Democracy Studies. He holds a master's degree from the University of Connecticut (2006), a teacher's and master's degree from the University of Mannheim (2007), and a Ph.D. from the University of Connecticut (2010). As the holder of the research chair, he sees himself as an ambassador to Canadian German relations. His research focuses on key challenges of Germany and Canada, and representative democracies more generally. These include (1) the effects of migration on political attitudes, (2) the populist tide that has swept the world, (3) transformations in the conduct of elections and the determinants of vote choice, and (4) unequal representation of various cohorts of the population including women, minorities, and youth. Throughout his academic career, Daniel has published 4 single-authored book, 1 edited volume, 1 textbook, and more than 130 articles in peer-reviewed journals.

Jean-Nicolas Bordeleau is a Ph.D. candidate in political science at the University of Ottawa. He holds a bachelor's degree in political science and psychology from the Royal Military College of Canada and a master's degree in political science from the Université de Montréal. His research focuses on the behavior of citizens within consolidated democracies. More specifically, his work explores the psychological and individual-level determinants of political behavior including electoral participation and vote choice. He is also active in the quantitative methodological literature, advancing research on experimental methods, and psychometric approaches in political science. Jean-Nicolas has published several research articles in peer-reviewed journals such as *International Journal*, *Democracy & Security*, and the *Election Law Journal*.

List of Figures

List of Tables

Introduction

1

Under what conditions do countries go to war? What is the influence of the 2008–2009 economic crisis on the vote share of radical right-wing parties in Western Europe? What type of people are the most likely to protest and partake in demonstrations? How has the urban squatters' movement developed in South Africa after Apartheid? What is the impact of the COVID-19 pandemic on conspiracy beliefs? There is hardly any field in the social sciences that asks as many research questions as political science. Questions scholars are interested in can be specific and reduced to one event (e.g., the development of the urban squatter's movement in South Africa post-Apartheid) or general and systemic such as the occurrence of war and peace. Whether general or specific, what all empirical research questions have in common is the necessity to use adequate research methods to answer them. For example, to effectively evaluate the influence of the economic downturn in 2008/2009 on the radical right-wing success in the elections preceding this crisis, we need data on the radical right-wing vote before and after the crisis, a clearly defined operationalization of the crisis, and data on confounding factors such as immigration, crime, and corruption. Through appropriate modeling techniques (i.e., multiple regression analysis on macro-level data), we can then assess the absolute and relative influence of the economic crisis on the radical right-wing vote share.

Research methods are the "bread and butter" of empirical political science. They are the tools that allow researchers to conduct research and detect empirical regularities, causal chains, and explanations of political and social phenomena. To use a practical analogy: a political scientist needs to have a toolkit of research methods at his or her disposal to build good empirical research in the same way as a mason must have certain tools to build a house. It is indispensable for a mason to not only have some rather simple tools (e.g., a hammer), but also some more

© The Author(s), under exclusive license to Springer Nature Switzerland AG 2023

D. Stockemer and J.-N. Bordeleau, *Quantitative Methods for the Social Sciences*,
Springer Texts in Political Science and International Relations,
https://doi.org/10.1007/978-3-031-34583-8_1

sophisticated tools such as a mixer or crane. The same applies for a political scientist. Ideally, he or she should have some easy tools (such as descriptive statistics or means testing) at his or her disposal, but also some more complex tools such as pooled time series analysis or maximum likelihood estimation. Having these tools allows political scientists to both conduct their own research and judge and evaluate other peoples' work. This book will provide a first simple tool kit in the areas of quantitative methods, survey research, and statistics.

There is one caveat in methods' training: research methods can hardly be learnt by just reading articles and books. Rather, they need to be learnt in an applied fashion. Similar to the mixture of theoretical and practical training a mason acquires during her apprenticeship, political science students should be introduced to method's training in a practical manner. In particular, this applies to quantitative methods and survey research. Aware that methods learning can only be fruitful if students learn to apply their theoretical skills in real-world scenarios, we have constructed this book on survey research and quantitative methods in a very practical fashion.

Through our own experiences as scholars and teachers of introductory courses into quantitative method, we have repeatedly learned that students only enjoy these classes if they see the applicability of the techniques they learn. This book follows the structure as laid down in Fig. 1.1; it is structured so that students learn various statistical techniques while using their own data. It does not require students to have taken prior methods' classes. To lay some theoretical groundwork, the first chapter starts with an introduction into the nuts and bolts of empirical social sciences (see Chap. 2). The book then shortly introduces students to the nuts and bolts of survey research (see Chap. 3). The following chapter then very briefly teaches students how they can construct and administer their own survey. At the end of Chap. 4, students also learn how to construct their own questionnaire. The fifth chapter, entitled 'Conducting a survey', instructs students on how to conduct a survey in the field. During this chapter, groups of students can also apply their knowledge and test their survey in an empirical setting by soliciting answers from peers. Chapters 6, 7, 8 and 9 are then dedicated to analyzing the survey. In more detail, students receive a thorough introduction to R, the statistical program we are using in this book to analyze data, in the first part of Chap. 6. The second part covers univariate statistics and graphical representations of the data. In Chap. 7, we introduce different forms of means testing. Chapter 8 is then dedicated to bivariate correlation and regression analysis. Finally, Chap. 9 covers multivariate regression analysis.

The book can be used as a self-teaching device. In this case, students should redo the exercises with the data provided. In a second step, they should conduct all the tests with other data they have at their disposal. The book is also the perfect accompanying textbook for an introductory class to survey research and statistics. In the latter case, there is a built-in semester-long group exercise, which enhances the learning process. In the semester long group work that follows the sequence of the book students are asked to conceive, conduct, and analyze their own survey. The survey that is analyzed throughout the textbook is a colloquial survey that

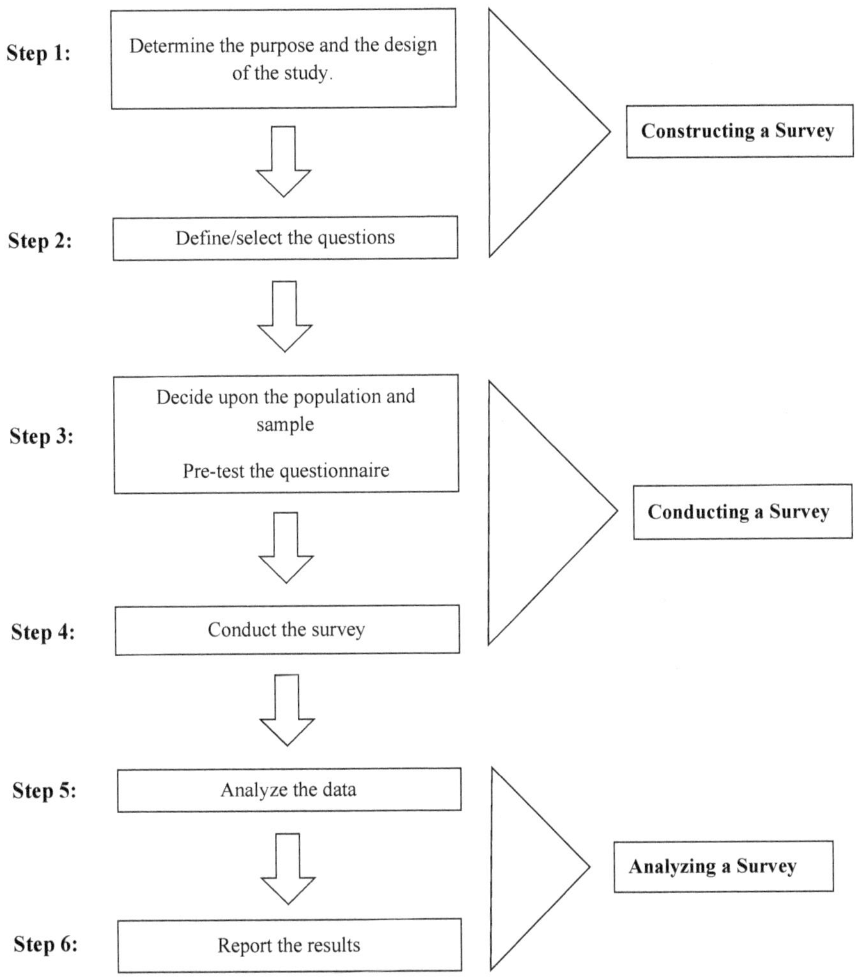
Step 1:
Determine the purpose and the design of the study.
Step 2:
Define/select the questions
Constructing a Survey
Step 3:
Decide upon the population and sample
Pre-test the questionnaire
Step 4:
Conduct the survey
Conducting a Survey
Step 5:
Analyze the data
Step 6:
Report the results
Analyzing a Survey

Fig. 1.1 Different steps in survey research

measures the amount of money students spend partying. It is an original survey including the original data, which a previous student group collected during their semester-long project. Using this "colloquial" survey, the students in this study group had lots of fun collecting and analyzing their "own" data, showing that learning statistics can (and should) be fun. We hope that the readers and users of this book experience the same joy in their first encounter with quantitative methods.

The Nuts and Bolts of Empirical Social Science

2

2.1 What is Empirical Research in the Social Sciences?

Regardless of its subdiscipline, empirical research in the social sciences tries to decipher how the world works around us. Be it development studies, economics, sociology, political science, or geography, just to name a few disciplines, researchers try to explain how some part of the world is structured. For example, political scientists may try to answer why some people vote while others abstain from casting a ballot. Scholars in developmental studies might look at the influence of foreign aid on economic growth in the receiving country. Researchers in the field of education studies might examine how the size of a school class impacts the learning outcomes of high school students, and economists might be interested in the effect of raising the minimum wage on job growth. Regardless of the discipline they are in, social science researchers try to explain the behavior of individuals such as voters, protesters, students, the behavior of groups such as political parties, companies, or social movement organizations, or the behavior of macro-level units such as countries.

While the tools taught in this book are applicable to all social science disciplines, we mainly cover examples from empirical political science, because this is the discipline in which we teach and research. In all social sciences and in political science, more generally, knowledge acquisition can be both normative and empirical. Normative political science asks the question of how the world ought to be. For example, normative democratic theorists quibble with the question of what a democracy should be. Is it an entity that allows free, fair, and regular elections, which in the democratic literature, is referred to as the "minimum definition of democracy"? (Boogards, 2007). Or must a country, in addition to having a fair electoral process, grant a variety of political rights (e.g., freedom of religion, the freedom of assembly), social rights (e.g., the right to health care and housing), and economic rights (e.g., the right to education or housing) to be "truly" democratic? This more encompassing definition is currently referred

© The Author(s), under exclusive license to Springer Nature Switzerland AG 2023

D. Stockemer and J.-N. Bordeleau, *Quantitative Methods for the Social Sciences*,
Springer Texts in Political Science and International Relations,
https://doi.org/10.1007/978-3-031-34583-8_2

to in the literature as the "maximum definition of democracy" (Beetham, 1999). While normative and empirically oriented research have fundamentally different goals, they are nevertheless complementary. To highlight, an empirical democracy researcher must have a benchmark when she defines and codes a country as a democracy or non-democracy. This benchmark can only be established through normative means. Normative political science must establish the "gold-standard" against which empirically oriented political scientists can test whether a country is a democracy or not.

As such, empirical political science is less interested in what a democracy should be, but rather how a democracy behaves in the real world. For instance, an empirical researcher could ask the following questions: Do democracies have greater gender representation in parliaments than non-democracies? Do democracies have less military spending than autocracies or hybrid regimes? Is the history curriculum in high schools different in democracies than in other regimes? Does a democracy spend more on social services than an autocracy? Answering these questions requires observation and empirical data. Whether it is collected at the individual level through interviews or surveys, at the meso-level through, for example, membership data of parties or social movements, or at the macro-level through government/international agencies or statistical offices, the collected data should be of high quality. Ideally, the measurement and data collection process of any study should be clearly laid down by the researcher, so that others can replicate the same study. After all, it is our goal to gain intersubjective knowledge. Intersubjective means that if two individuals would engage in the same data collection process and would conduct the same empirical study, their results would be analogous. To be as intersubjective or "facts based" as possible empirical political science should abide by the following criteria:

Falsifiability: the falsifiability paradigm implies that statements or hypotheses can be proven or refuted. For example, the statement that democracies do not go to war with each other can be tested empirically. After defining what a war and democracy is, we can get data that fits our definition for a country's regime type from a trusted source like the *Polity IV* data verse and data for conflict/war from another high-quality source such as the *UCDP/PRIO Armed Conflict* dataset. In second step, we can then use statistics to test whether the statement that democracies refrain from engaging in warfare with each other is true or not.[1,2]

Transmissibility: the process through which we can achieve the transmissibility of research findings is called replication. Replication refers to the process by which prior findings can be retested. Retesting can involve either the same data or new

[1] The PolityIV database adheres to rather minimal definition of democracy. In essence, the database gauges the fairness, competitiveness of the elections and the electoral process on a scale from −10 to +10. −10 describes the "worst" autocracy, while 10 describes a country full respects free, fair, and competitive elections.

[2] The UCDP/ PRIO Armed Conflict Dataset defines minor wars by a death toll between 25 and 1000 people and major wars by a death toll of 1000 people and above (see Gleditsch, 2002).

data from the same empirical referents. For instance, the "law-like" statement that democracies do not go to war with each other could be retested every five years with the most recent data from *Polity IV* and the *UCDP/PRIO Armed Conflict* dataset covering these five years to see if it still holds. Replication involves high scientific standards; it is only possible to replicate a study if the data collection, the data source, and the analytical tools are clearly explained and laid down in any piece of research. The replicator should then also use these same data and methods for her replication study.

Cumulative nature of knowledge: Empirical scientific knowledge is cumulative. This entails that substantive findings and research methods are based upon prior knowledge. In short, researchers do not start from scratch or intuition when engaging in a research project. Rather, they try to confirm, amend, broaden, or build upon prior research and knowledge. For example, the statement that democracies avoid war with each other had been confirmed and reconfirmed many times in the 1980s, 1990s, and 2000s (see Russett, 1994; De Mesquita et al., 1999). After confirming that the *Democratic Peace Theory* in its initial form is solid, researchers tried to broaden the democratic peace paradigm and examined, for example, if countries that share the same economic system (e.g., neoliberalism) also do not go to war with each other. Yet, for the latter relationship, tests and retests have shown that the empirical linkage for the economic system's peace is less strong than the democratic peace statement (Chandler, 2010). The same applies to another possible expansion, which looks at if democracies, in general, are less likely to go to war than non-democracies. Here again the empirical evidence is negative, or inconclusive at best (Daase, 2006; Mansfield & Snyder, 2007).

Generalizability: In empirical social science, we are interested in general rather than specific explanations; we are interested in boundaries or limitations of empirical statements. Does an empirical statement only apply to a single case (e.g., does it only explain why the US and Canada have never gone to war), or can it be generalized to explain many cases (e.g., does it explain why all possible dyads of democracies don't go to war). In other words, if it can be generalized, does the democratic peace paradigm apply to all democracies, only to neoliberal democracies, and does it apply across all (normative) definitions of democracies, as well as all time periods? Stated differently, we are interested in the number of cases in which the statement is applicable. Of course, the broader the applicability of an explanation, the more weight it carries. In political science, the Democratic Peace Theory is among the theories with the broadest applicability. While there are some questionable cases of conflict between states such as the conflict between Turkey and Greece over Cyprus in 1974, there has, so far, been no case that clearly disproves the Democratic Peace Theory. In fact, the Democratic Peace Theory is one of the few law-like rules in political science.

2.2 Qualitative and Quantitative Research

In the social sciences, we distinguish two large strands of research: quantitative and qualitative research. The major difference between these two research traditions is the number of observations. Research that involves few observations (e.g., one, two, or three individuals or countries) is generally referred to as qualitative. Such research requires an in-depth examination of the cases at hand. In contrast, works that includes hundreds, thousands, or even hundreds of thousands of observations are generally called quantitative research. Quantitative research works with statistics or numbers that allow researchers to quantify the world. In the twenty-first century, statistics are nearly everywhere. In our daily lives, we encounter statistics in approval ratings of TV shows, in COVID-19 pandemic coverage, the measurement of consumer preferences, weather forecasts, and betting odds, just to name a few examples. In social and political science research, statistics are the bread and butter of much scientific inquiry; statistics help us make sense of the world around us. For instance, in the political realm, we might gauge turnout rates as a measurement of the percentage of citizens that cast their ballot during an election. In economics, some of the most important indicators about the state of the economy are monthly growth rates and consumer price indexes. In the field of education, the average grade of a student from a specific school gives an indication of the school's quality.

By using statistics, quantitative methods not only allow us to numerically describe phenomena, but they also help us determine relationships between two or more variables. Examples of these relationships are multifold. For example, in the field of political science, statistics and quantitative methods have allowed us to detect that citizens who have a higher socio-economic status (SES) are more likely to vote than individuals with a lower socio-economic status (Milligan et al., 2004). In the field of economics, researchers have established with the help of quantitative analysis that low levels of corruption foster economic growth (Mo, 2001). And in education research, there is near consensus in the quantitative research tradition that students from racially segregated areas and poor inner-city schools, on average, perform less strongly in college entry exams than students from rich, white neighborhoods (Rumberger & Palardy, 2005).

Quantitative research is the primary tool to establish empirical relationships. However, it is less well-suited to explain the constituents or causal mechanisms behind a statistical relationship. To highlight, quantitative research can illustrate that individuals with low education levels and below average income are less likely to vote compared to highly educated and rich citizens. Yet, it is less suitable to explain the reasons for their abstentions. Do they not feel represented? Are they fed up with how the system works? Do they not have the information and knowledge necessary to vote? Similarly, quantitative research robustly tells us that students in racially segregated areas tend to perform less strongly than students in predominantly white and wealthy neighborhoods. However, it does not tell us, how the disadvantaged students feel about these inequalities and what they think can be done to reverse them. Are they enraged or fed up with the political regime and the politicians that represent it? Questions like these are better answered by qualitative research. The qualitative researcher wants to interpret the observational data (i.e., the fact that low SES individual have a higher likelihood to vote) and wants to grasp the opinions and attitudes of study subjects (i.e., how minority students feel in disadvantaged areas, how they think the system perpetuates these inequalities, and under what circumstances they are ready to protest). To gather this in-depth information, the qualitative researcher uses different techniques than the quantitative researchers. She needs research tools to tap into the opinions, perceptions, and feelings of study subjects. Tools appropriate for these inquiries are ethnographic methods including qualitative interviewing, participant observations, and the study of focus groups. These tools help us understand how individuals live, act, think, and feel in their natural setting and give meaning to quantitative findings.

In addition to allowing us to decipher meaning behind quantitative relationships, qualitative research techniques are an important tool in theory building. In fact, many research findings originate in qualitative research and are tested in a later stage in a quantitative large-N study. To take a classic in social sciences, Theda Skocpol offers in her seminal work "States and Social Revolutions: A comparative Analysis of Social Revolutions in Russia, France and China (1979)", an explanation for the occurrence of three important revolutions in the modern world, the French Revolution in 1789, the Chinese Revolution in 1911, and the Russian Revolution in 1917. Through historical analysis, Skocpol identifies three conditions for a revolution to happen: (1) a profound state crisis, (2) the emergence of a dominant class outside of the ruling elites, and (3) a state of severe economic and/ or security crisis. Skocpol's book is an important exercise in theory building. She identifies three causal conditions; conditions that are quantifiable and that can be tested for other or all revolutions. By testing whether a profound state crisis, the emergence of a dominant class outside of the ruling elites and a state of crisis explain other or all revolutions, quantitative researchers can establish the boundary conditions of Skocpol's theory.

It is also important to note that not all research is quantifiable. Some phenomena such as individual identities or ideologies are difficult to reduce to numbers: What are ethnic identities, religious identities or regional identities? Often these critical concepts are not only difficult to identify, but frequently also difficult to grasp empirically. For example, to understand what the regional francophone identity of Quebecers is, we need to know the historical, social, and political context of the province and the fact that the province is surrounded by English speakers. To get a complete grasp of this regional identity, we, ideally, also must retrace the recent development that more and more English is spoken in the major cities of Québec such as Montréal, particularly in the business world. These complexities are hard to reduce to numbers and need to be studied in-depth. For other events, there are just not enough observations to quantify them. For example, the Cold War is a unique event, an event that organized and shaped the world for 45 years in the twentieth century. Nearly, by definition this even is important and needs to be studied in depth. Other events, like World War I and World War II, are for sure a subset of wars. However, these two wars have been so important for world history, that, nearly by definition, they require in depth study, as well. Both wars have shaped who we as individuals are (regardless where we live), what we think, how we act, and what we do. Hence, any bits of additional knowledge we acquire from these events not only help us understand the past, but also helps us move forward in the future.

Quantitative and qualitative methods are complimentary; students of the social sciences should master both techniques. However, it is hardly possible to do a thorough introduction into both. This book is about survey research, quantitative research tools, and statistics. It will teach you how to draft, conduct, and analyze a survey. However, before delving into the nuts and bolts of data analysis we need to know what theories, hypotheses, concepts, and variables are. The next section will give you a short overview of these building blocks in social research.

2.3 Theories, Concepts, Variables, and Hypotheses

2.3.1 Theories

We have already learnt that social science research is cumulative. We build current knowledge on prior knowledge. Normally, we summarize our prior knowledge in theories, which are parsimonious or simplified explanations of how the world works. As such, a theory summarizes established knowledge in a specific field of study. Because the world around us is dynamic, a theory in the social sciences is never a deterministic statement. Rather it is open to revisions and amendments.[3] Theories can cover the micro-, meso-, and macro-levels. Below are three powerful social sciences theories

Example of a micro-level theory: relative deprivation

Relative deprivation is a powerful individual level theory to explain and predict citizens' participation in social movement activities. Relative deprivation starts with the premise that individuals do not protest, when they are happy with their lives. Rather grievance theorists (e.g., Gurr, 1970; Runciman, 1966) see a discrepancy between value expectation and value capabilities as the root cause for protest activity. For example, according to Gurr (1970) individuals normally have no incentive to protest and voice their dissatisfaction if they are content with their daily lives. However, a deteriorating economic, social, or political situation can trigger frustrations, whether or real or perceived; the higher these frustrations are the higher the likelihood that somebody will protest.

[3] The idea behind parsimony is that scientists should rely on as few explanatory factors as possible while retaining a theory's generalizability.

Example of a meso-level theory: the iron law of oligarchy

The iron law of oligarchy is a political meso-level theory developed by German sociologist Robert Michels. His main argument is that over time all social groups, including trade unions and political parties, will develop hierarchical power structures or oligarchic tendencies. Stated differently, in any organization a "leadership class" consisting of paid administrators, spokespersons, societal elites, and organizers will prevail and centralize its power. And with power comes the possibility to control the laws and procedures of the organization, the information it communicates and the power to reward faithful members; all these tendencies are accelerated by apathetic masses, which will allow elites to hierarchize an organization faster (see Michels, 1915).

Example of a macro-level theory: the democratic peace theory

As discussed earlier in this chapter, a famous example of a macro-level theory is the so-called democratic peace theory, which dates back to Kant's treatise on Perpetual Peace (1795). The theory states that democracies will not go to war with each other. It explicitly tackles the behavior of some type of state (i.e., democracies) and has only applicability at the macro-level.

Theory development is an iterative process. Because the world around us is dynamic (what is true today might no longer be true tomorrow), a theory must be perpetually tested and retested against reality. The more it is confirmed across time and space, the more it is robust. Theory building is a reiterative and lengthy process. Sometimes it takes years, if not decades to build and construct a theory. A famous example of how a theory can develop and refine is the simple rational choice theory of voting. In his 1957 famous book, *An Economic Theory of Democracy*, Anthony Downs tries to explain why some people vote, whereas others abstain from casting their ballots. Using a simple rational choice explanation, he concludes that voting is a "rational act" if the benefits of voting surpass the costs. To operationalize his theory, he defines the benefits of voting by the probability that an individual vote counts. The costs include the physical costs of actually leaving one's house and casting a ballot, as well as the ideational costs of gathering the necessary information to cast an educated ballot. While Downs finds his theory logical, he intuitively finds that there is something wrong with it. That is, the theory would predict that in the overall majority of cases citizens should not vote, because in almost every case the probability that an individual's vote will count is close to 0. Hence, the costs of voting surpass the benefits of voting for nearly every individual. However, Downs finds that in the majority people still vote, but does not have an immediate answer for this paradox of voting.

More than 10 years later, in a reformulation of Downs' theory, Riker and Ordeshook (1968) resolved Downs' paradox by adding an important component to Downs' model; the intangible benefits. According to the authors, the benefits of voting are not reduced to pure materialistic evaluations (i.e., the chance that a person's vote counts), but also to some non-materialistic benefits such as citizens' willingness to support democracy or the democratic system. Adding this additional component makes Downs' theory more realistic and in tune with reality. On the negative side, adding non-material benefits makes Downs' theory less parsimonious. However, all researchers would probably agree that this sacrifice of parsimony is more than compensated for by the higher empirical applicability of the theory. Therefore, in this case the more complex theory is preferential to the more parsimonious theory. More generally, a theory should be as simple or parsimonious as possible and as complex as necessary.

2.3.2 Concepts

Theories are abstractions of objects, objects' properties, or behavioral phenomena. Any theory normally consists of at least two concepts, which define a theory's content and attributes. For example, the Democratic Peace Theory consists of the two concepts: democracy and war. Some concepts are concise (e.g., wealth, education, women's representation) and easier to measure, whereas other concepts are abstract (democracy, equal opportunity, human rights, social mobility, political culture) and more difficult to gauge. Whether abstract or precise, concepts provide a common language for political science. For sure, researchers might disagree about the precise (normative) definition of a concept. Nevertheless, they agree about its meaning. For example, if we talk about democracy, there is common understanding that we talk about a specific regime type that allows free and fair and elections and some other freedoms. Nevertheless, there might be disagreement about the precise definition of the concept in question; in the case of democracy disagreement might revolve the following questions: do we only look at elections, do we include political rights, social rights, economic rights or all of the above? To avoid any confusion, researchers must be precise when defining the meaning of a concept. In particular, this applies for contested concepts such as the aforementioned democracy. As already mentioned, for some scholars, the existence of parties, free and fair elections and a reasonable participation by the population might be enough to classify a country as a democracy. For others, a country must have legally enforced guarantees for freedoms of speech, press, and religion as well as must guarantee social and economic rights. It can be either a normative or a practical question, or both whether one or the other classification is more appropriate. It might also be a question of the specific research topic or research question whether one or the other definition is more appropriate. Yet, whatever definition she chooses, a researcher must clearly identify and justify the choice of her definition, so that the reader of a published work can judge the appropriateness of the chosen definition.

It is also worth noting that the meaning of concepts can also change over time. To take again the example of democracy. Democracy 2000 years ago had a different meaning than democracy today. In the Greek city states (e.g., Athens), democracy was a system of direct decision making, in which all men above a certain income threshold convened on a regular basis to decide upon important matters such as international treaties, peace and war, as well as taxation. Women, servants, slaves, and poor citizens were not allowed to participate in these direct assemblies. Today, more than 2000 years after the Greek city states, we commonly refer to democracy as a representative form of government, in which we elect members to parliament. In the elected assembly, these elected politicians should then represent the citizens that mandated them to govern. Despite the contention of how many political, civic, and social rights are necessary to consider a country a democracy, there is nevertheless agreement among academics and practitioners today that the Greek definition of democracy is outdated. In the twenty-first century, no serious academic would disagree that suffrage must be universal, each vote must count equally, and elections must be free and fair and must occur on a regular basis such as in a 4- or 5-year interval.

2.3.3 Variables

A variable refers to properties or attributes of a concept that can be measured in some way or another: in short, a variable is a measurable version of a concept. The process to transform a concept into a variable is called operationalization. To take an example, age is a variable, but the answer to the question how old you are, is a variable. Some concepts in political or social science are rather easy to measure. For instance, on the individual level, somebody's education level can be measured by the overall years of schooling somebody has achieved or by the highest degree somebody has obtained. On the macro-level, women's representation in parliament can be easily measured by the percentage of seats in the elected assembly which are occupied by women. Other concepts, such as someone's political ideology on the individual level or democracy on the macro-level are more difficult to measure. For example, operationalizations of political ideology range from the party one identifies with, to answers to survey questions about moral issues such as abortion or same sex marriage, to questions about whether somebody prefers more welfare state spending and higher taxes, or less welfare state spending and lower taxes. For democracy, as already discussed, there is not only discussion of the precise definition of democracy, but also on how to measure different regime types. For example, there is disagreement in the academic literature if we should adopt a dichotomous definition that distinguishes a democracy from a non-democracy (Przeworski et al., 1996), a distinction in democracy, hybrid regime or autocracy (Bollen, 1990), or if we should use a graded measure that is democracy is not a question of kind, but of degree, and the gradation should capture sometimes partial processes of democratic institutions in many countries (Elkins, 2000).

When measuring a concept, it is important that such a concept has high content validity; there should be a high degree of convergence between the measure and the concept. In other words, a high content validity is achieved if a measure represents all facets of a given concept. To highlight how this convergence can be achieved, we use one famous definition of democracy, Dahl's Polyarchy. Polyarchy, according to Dahl, is a form of representative democracy characterized by a particular set of political institutions. These include elected officials,

Table 2.1 Measuring Dahl's polyarchy

Components of democracy	Country 1	Country 2	Country 3
Elected officials have control over government decisions	x	–	x
Free, fair, and frequent elections	x	–	x
Universal suffrage	x	–	x
Right to run for office for all citizens	x	–	x
Freedom of expression	x	–	–
Alternative sources of information	x	–	–
Right to form and join autonomous political organizations	x	–	x
Polyarchy	Yes	No	No

free and fair elections, inclusive suffrage, the right to run for office, freedom of expression, alternative information, and associational autonomy (see Dahl, 1973). To achieve high content validity, any measurement of polyarchy must include the seven dimensions of democracy; that is any of these seven dimensions must be explicitly measured. Sometimes a conceptual definition predisposes researchers to use one operationalization of a concept over another one. In Dahl's classification, the respect of the 7 features is a minimum standard for democracy; that is why, his concept of polyarchy is best operationalized dichotomously. That is a country is a polyarchy if it respects all of the seven features and is not if it doesn't (i.e., it is enough to not qualify a country a country as a non-democracy if one of the features is not respected). Table 2.1 graphically displays this logic. Only country 1 respects all features of a polyarchy and can be classified as such. Countries 2 and 3 violate some or all of these minimum conditions of polyarchy, and hence must be coded as non-democracies.

Achieving high content validity is not always easy. Some concepts are difficult to measure. Take the concept of political corruption. Political corruption, or the private (mis-)use of public funds for illegitimate private gains, happens behind closed doors without the supervision of the public. Nearly by definition this entails that nearly all proxy variables to measure corruption are imperfect. There are at least three ways to measure corruption:

1. Large international efforts compiled by international organizations such as the World Bank or Transparency International try to track corruption in the public sector around the globe. For example, Corruption Perceptions Index (CPI) focuses on corruption in the public sector. It uses expert surveys with country experts inside and outside the country under scrutiny on, among others, bribery of public officials, kickbacks in public procurement, embezzlement of public funds, and the strength and effectiveness of public sector anti-corruption efforts. It then creates a combined measure from these surveys.
2. National agencies in several (western) countries track data on the number of federal, state, and local government officials prosecuted and convicted for corruption crimes.

3. International public opinion surveys (e.g., the World Value Survey) ask citizens about their own experience with corruption (e.g., if they have paid a bribe to an official for any public service received within the past 12 months).

Any of these three measurements is potentially problematic. First, perception indexes based on interviews/ surveys with country experts can be deceiving, as there is no hard evidence to back up claims of high or low corruption, even if these assessments come from so-called experts. However, the hard evidence can be deceiving as well. Are many corruption charges and indictments a sign of high or low corruption? They might be a sign of high corruption, as it shows corruption is widespread; a certain percentage of the officials in the public sector engage in the exchange of public goods for private promotion. Yet, many cases treated in court might also be a sign of low corruption. It might show that the system works, as it cracks down on corrupted officials. The third measure, citizens' personal experience with corruption is suboptimal, as well. Given that corruption is a shameful act, survey participants might not admit that they have participated in fraudulent activities. They might also fear repercussions by the public or government officials if they admit being part of a corrupt network. Finally, it might not be rational to admit corruption, particularly if you are one of the beneficiaries of it.

In particular, for difficult to measure concepts such as corruption, it might be advisable to cross-validate any imperfect proxy with another measure. In other words, different measures must resemble each other if they tap into the same concept. If this is the case we speak of high construct validity, and it is possibly safe to use one proxy, or even better create a conjoint index of the proxy variables in question. If this is not the case, then there is a problem with one or several measurements, something, the researcher should assess in detail. One way to measure whether two measurements of the same variable are strongly related to each other is through correlation analysis (see Chap. 8 of this book).

Sometimes it is not only difficult to achieve high operational validity of difficult concepts such as corruption, but sometimes also for seemingly simple concepts such as voting or casting a ballot for a radical right-wing party. In answering a survey, individuals might pretend they have voted or cast a ballot for a mainstream party to make-believe that they abide by the societal norms. Yet, it is very difficult to detect the type of individuals, who either deliberately or non-deliberately answer a survey question incorrectly (for a broader discussion of biases in survey research see Sect. 5.2.)

2.3.3.1 Types of Variables

In empirical research, we distinguish two main types of variables: dependent variables and independent variables.

Dependent Variable: the dependent variable is the variable the researcher is trying to explain. It is the primary variable of interest and depends on other variables (so-called independent variables). In quantitative studies, the dependent variable has the notation y.

Independent Variable: Independent variables are hypothesized to explain variation in the dependent variable. Because they should explain variation or changes in the dependent variable, independent variables are sometimes also called explanatory variables (as they should explain the dependent variable). In quantitative studies, the independent variable has the notation x.

We use another famous theory, modernization theory to explain the difference between independent and dependent variable. In essence, modernization theory states that countries with a higher degree of development are more likely to be democratic (Lipset, 1959). In this example, the dependent variable is regime type (however measured). The independent variable is a country's level of development, which could, for instance, be measured by a country's GDP per capita.

In the academic literature, independent variables that are not the focus of the study, but which might also have an influence on the dependent variable, are sometimes referred to as control variables. To take an example from the turnout literature: a researcher might be interested in the relationship between electoral competitiveness and voter turnout. Electoral competitiveness is the independent variable and turnout is the dependent variable. However, turnout rates in countries or electoral districts are not only dependent on the competitiveness of the election (which is often operationalized by the difference in votes between the winner and the runner-up), but also by a host of other factors including compulsory voting, the electoral system type, corruption or income inequalities, to name a few factors. These other independent variables must also be accounted for and included in the study. In fact, researchers can only test the "real impact" of electoral competitiveness on turnout, if they also take these other factors into consideration.

2.3.4 Hypothesis

A hypothesis is a tentative, provisional, or unconfirmed statement derived from theory that can (in principle) be either verified or falsified. It explicitly states the expected relationship between an independent and dependent variable. Hypotheses must be empirically testable statements that can cover any level of analysis. In fact, a good hypothesis should specify the types of political and level of analysis to which the hypothesis will apply (see also Table 2.2).

Macro-level: An example of a macro-level hypothesis derived from modernization theory would be: The more highly developed a country is, the more likely it is a democracy.

Table 2.2 Examples of good and bad hypotheses

Wrong	Right
Democracy is the best form of government	The more democratic a country is, the better its government performance will be
The cause of civil war is economic upheaval	The more there is economic upheaval, the more likely a country will experience civil war
Raising the US minimum wage will affect job growth	Raising the minimum wage will create more jobs (positive relationship) Raising the minimum wage will cut jobs (negative relationship)

Meso-level: An example of a meso-level hypothesis derived from the iron law of oligarchy would be: The longer a political or social organization is in existence, the more hierarchical are its power structures.

Micro-level: An example of micro-level hypothesis derived from the resource theory of voting would be: The higher somebody's level of education the more likely this person will vote.

Scientific hypotheses are always stated in the following form:

the more [independent variable] the more [dependent variable] **or** the more [independent variable] the less [dependent variable].

When researchers formulate hypotheses, they make three explicit statements:

1. X and Y covary. This implies that there is variation in the independent and dependent variable and that at least some of the variation in the dependent variable is explained by variation in the independent variable
2. Change in X precedes change in Y. By definition a change in the independent variable can only trigger a change in the dependent variable if this change happens before the change in the dependent variable.
3. The effect of the independent variable on the dependent variable is not coincidental or spurious (which means explained by other factors), but direct.

To provide an example, the resource theory of voting states that individuals with higher socio-economic status (SES) are more likely to vote. From this theory, we can derive the micro-level hypothesis that the more educated a citizen is the higher the chance that she will cast a ballot. To be able to test this hypothesis, we operationalize SES by a person's years of full-time schooling and voting by a survey question asking whether somebody voted or not in the last national election. By formulating this hypothesis, we make the implicit assumption that there is variation in the overall years of schooling and variation in voting. We also explicitly state that the causal explanatory chain goes from education to voting (i.e., that education precedes voting). Finally, we expect that changes in somebody's education trigger changes in somebody's likelihood to vote (i.e., we expect the relationship to not be spurious). While for the resource hypothesis, there is probably consensus that the causal chain goes from more education to a higher likelihood to vote, and not the other way around, the same does not apply to all empirical relationships. Rather, in political science we do not always have a one-directional relationship. For example, regarding the modernization hypothesis, there is some debate in the scholarly community surrounding whether it is development that triggers the installation of democracy, or if it is that democracy triggers robust economic development more than any other regime type. There are statistical methods to treat cases of reversed causation such as structural equation modeling. Because of the advanced nature of these techniques, we will not cover these techniques in this book. Nevertheless, what is important to take away from this discussion is that students of political science, and the social sciences more generally, must think carefully about the direction of cause and effect before they formulate a hypothesis.

It is also important that students know the difference between an alternative hypothesis and a null hypothesis. The alternative hypothesis, sometimes also called research hypothesis, is the hypothesis you are going to test. The null hypothesis is the rival hypothesis - it assumes that there is no association between the independent and dependent variables. To give an example derived from the Iron Law of Oligarchy: a researcher wanting to test this theory could postulate the hypothesis that "the longer a political organization is in existence, the more hierarchical it will get". In social science jargon, this hypothesis is called the alternative hypothesis. The corresponding null-hypothesis would be that length of existence of a political organization and its hierarchical structure are unrelated.

2.4 The Quantitative Research Process

The quantitative research is process is deductive (see Fig. 2.1). It is theory driven; it starts and ends with theory. Before the start of any research project, students of political science must know the relevant literatures. They must know the dominant theories and explanations of the phenomenon they want to study and identify controversies and holes or gaps in knowledge. The existing theory will then guide them to formulate some hypotheses that will ideally try to resolve some of the controversies or fill one or several gaps in knowledge. Quantitative research might also test existing theories with new quantitative data, establish the boundaries or limitations of a theory, or establish the conditions under which a theory applies. Whatever its purpose, good research starts with a theoretically derived research question and hypothesis. Ideally, the research question should address a politically/ relevant and important topic and make a potential theoretical contribution to the literature (it should potentially add to, alter, change, or refute the existing theory). The hypothesis should clearly identify the independent and dependent variable. It should be a plausible statement of how the researcher thinks that the independent variable behaves toward the dependent variable. In addition, the researcher must also identify potential control variables. In the next step, the researcher must think about how to measure independent-, dependent-, and control variables. When operationalizing her variables, she must ensure that there is high content validity between the numerical representation and the conceptional definition of any given concept. After having decided how to measure the variables, the researcher has to think about sampling. In other words, which empirical referents will she use to test her hypothesis? Measurement and sampling are often done concurrently, because the empirical referents, which the researchers studies, might predispose her to use one operationalization of an indicator over another. Sometimes, also practical considerations such as the existence of empirical data determine the measurement of variables and the number and type of observations studied. Once, the researcher has her data, she can then conduct the appropriate statistical tests to evaluate her research question and hypothesis. The results of her study will then ideally have an influence on theory.

Let us explain Fig. 2.1 with a concrete example. We assume that a researcher is interested in individuals' participation in demonstrations. Reading the literature, she finds two dominant theories. On the one hand, the resource theory of political action states that the more resources individuals have in the form of civic skills, network connections, time, and money, the more likely they are to engage in collective political activities including partaking in demonstrations. One the other hand, the relative deprivation approach states that individuals must be frustrated with their economic, social, and political situation. The more they see a gap between value expectations and value capabilities, the more likely they are going to protest. Implicit in the second argument is that individuals in the bottom echelon of society such as the unemployed, those, who struggle economically, or those,

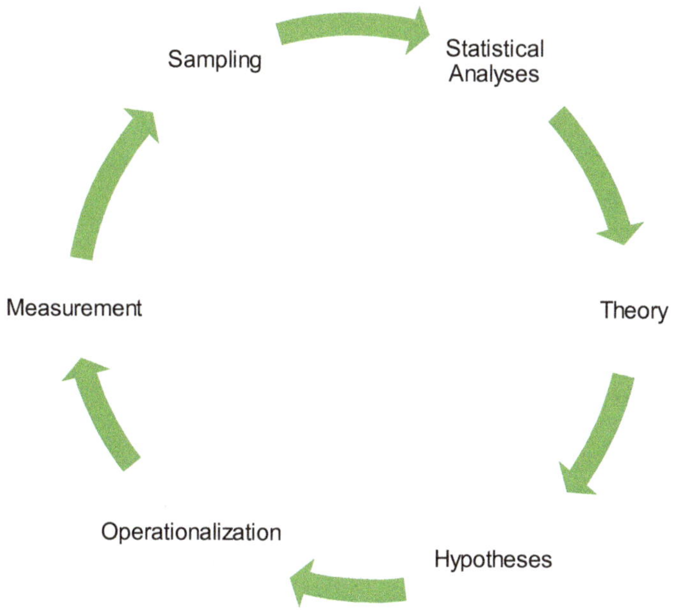

Fig. 2.1 Display of the quantitative research process. Adapted from Walsh and Ollenburger (2001)

who are deprived of equal chances in society such as minorities are more likely to demonstrate. Having identified this controversy, the researcher asks himself which, if either, of the two competing theories is more correct. Because, the researcher does not know, a priori, which of the two theories is more likely to apply she formulates two competing hypotheses:

Hypothesis 1: The higher somebody's SES the higher somebody's likelihood to partake in a demonstration.

Hypothesis 2: The higher somebody's dissatisfaction with her daily life, the higher the likelihood that this person will demonstrate.

Having formulated her hypotheses, the researcher has to identify other potentially relevant variables that could explain one's decision to partake in a demonstration. From the academic literature on protest, she identifies gender, age, political socialization, and place of residency as other potentially relevant variables which she must also include/control for in her study. Once the hypotheses are formulated and control variables identified, the researcher then must determine the measurement of the main variables of interest and for the control variables before finding an appropriate study sample. To measure the first independent variable, a person's SES, the researcher decides to employ two very well-known proxy variables, education, and income. For the second, independent variable, she thinks that the survey question "how satisfied are you with your daily life" captures individuals' levels of frustrations pretty well. The dependent variable, partaking in a demonstration, could be measured by a survey question asking whether somebody has demonstrated within the past year. Because the researcher finds that the European Social Survey (ESS) asks all these questions using a representative sample of individuals in about 20 European countries, she uses this sample as the study object or data source. She then engages in appropriate statistical techniques to gauge the influence of her two main variables of interest on the dependent variable.

References

Beetham, D. (1999). *Democracy and human rights*. Polity.

Bogaards, M. (2007). Measuring democracy through election outcomes: A critique with African data. *Comparative Political Studies, 40*(10), 1211–1237.

Bollen, K. A. (1990). Political democracy: Conceptual and measurement traps. *Studies in Comparative International Development (SCID), 25*(1), 7–24.

Chandler, D. (2010). The uncritical critique of 'liberal peace.' *Review of International Studies, 36*(1), 137–155.

Daase, C. (2006). Democratic peace—democratic War: Three Reasons why democracies are war-prone. In *Democratic wars* (pp. 74–89). Palgrave Macmillan UK.

Dahl, R. A. (1973). *Polyarchy: Participation and opposition*. Yale University Press.

De Mesquita, B. B., Morrow, J. D., Siverson, R. M., & Smith, A. (1999). An institutional explanation of the democratic peace. *American Political Science Review, 93*(4), 791–807.

Elkins, Z. (2000). Gradations of democracy? Empirical tests of alternative conceptualizations. *American Journal of Political Science, 44*(2), 293–300.

Gleditsch, N. E. (2002). Armed conflict 1946–2001: A new dataset. *Journal of Peace Research, 39*(5), 615–637.
Gurr, T. R. (1970). *Why men rebel.* Princeton University Press.
Kant, I. (1795) [2011]. *Zum ewigen Frieden* (3rd ed.). Akademie Verlag.
Lipset, S. M. (1959). Some social requisites of democracy: Economic development and political legitimacy. *American Political Science Review, 53*(1), 69–105.
Mansfield, E. D., & Snyder, J. (2007). *Electing to fight: Why emerging democracies go to war.* MIT Press.
Michels, R. (1915). *Political parties: A sociological study of the oligarchical tendencies of modern democracy.* The Free Press.
Milligan, K., Moretti, E., & Oreopoulos, P. (2004). Does education improve citizenship? Evidence from the United States and the United Kingdom. *Journal of Public Economics, 88*(9), 1667–1695.
Mo, P. H. (2001). Corruption and economic growth. *Journal of Comparative Economics, 29*(1), 66–79.
Przeworski, A., Alvarez, M., Cheibub, J. A., & Linongi, F. (1996). What makes democracies endure? *Journal of Democracy, 7*(1), 39–55.
Riker, W. H., & Ordeshook, P. C. (1968). A theory of the calculus of voting. *American Political Science Review, 62*(1), 25–42.
Rumberger, R. W., & Palardy, G. J. (2005). Does segregation still matter? The impact of student composition on academic achievement in high school. *Teachers College Record, 107*(9), 1999.
Runciman, W. G. (1966). *Relative deprivation and social injustice. A study of attitudes to social inequality in twentieth century England.* Routledge and Kcagan Paul.
Russett, B. (1994). *Grasping the democratic peace: Principles for a post-cold war world.* Princeton University Press.
Skocpol, T. (1979). *States and social revolutions: A comparative analysis of France, Russia and China.* Cambridge University Press.
Walsh, A., & Ollenburger, J. C. (2001). *Essential statistics for the social and behavioral sciences: A conceptual approach.* Prentice Hall.

Further Readings

Research Design

Creswell, J. W., & Creswell, J. D. (2017). *Research design: Qualitative, quantitative, and mixed methods approaches.* Sage publications.

Nice introduction into the two main research traditions qualitative and quantitative research. The book also covers mixed methods' approaches (approaches that combine qualitative and quantitative methods).

McNabb, D. E. (2015). *Research methods for political science: Quantitative and qualitative methods.* Routledge (Chap. 7).

Nice introduction into the nuts and bolts of quantitative methods. Introduces basic concepts such as reliability and validity, as well as discusses different types of statistics (i.e., inferential statistics).

Shively, W. P. (2016). *The craft of political research.* Routledge.

Precise and holistic introduction into the quantitative research process.

Theories and Hypotheses

Brians, C. L., Willnat, L., Manheim, J., & Rich, R. (2016). *Empirical political analysis.* Routledge (Chaps. 2, 4, 5).

Comprehensive introduction into theories, hypothesis testing and operationalization of variables.

Qualitative research

Elster, J. (1989). *Nuts and bolts for the social sciences.* Cambridge University Press.

A nice introduction into causal explanations and causal mechanisms. The book explains what causal mechanisms are and what research steps the researcher can conduct to detect them.

Gerring, J. (2004). What is a case study and what is it good for? *American Political Science Review, 98*(2), 341–354.

A nice introduction on what a case study is, what is good for in political science, and what different types of case studies exist.

Lijphart, A. (1971). Comparative politics and the comparative method. *American Political Science Review, 65*(3), 682–693.

Seminal work on the comparative case study. Explains what a comparative case study is, how it relates to the field of comparative politics, and how to conduct a comparative case study.

A Short Introduction to Survey Research

3

3.1 What is Survey Research?

Survey research has become a major, if not the main, technique to gather information about individuals of all sorts. To name a few examples:

- **Customer surveys** ask individuals about their purchasing habits or their satisfaction with a product or service. Such surveys can reveal consumer habits and inform marketing strategies by companies.
- **Attitudinal surveys** poll participants on social, economic, or cultural attitudes. These surveys are important for researchers and policy makers as they allow us to detect cultural values, political attitudes, and social preferences.
- **Election surveys** ask citizens about their voting habits. As such they can, for example, influence campaign strategies by parties.

Regardless of its type, survey research involves the systematic collection of information from individuals using standardized procedures. When conducting survey research, the researcher normally uses a (random or representative) sample from the population she wants to study and asks the survey respondents one or several questions about attitudes, perceptions, or behaviors. In the ideal case, she wants to produce a set of data on a given phenomenon that captures the studied concept as well as relevant independent variables. She also wants to have a sample that describes the population she wants to study fairly well (Fowler, 2009: 1). To provide a concrete example, if a researcher wants to gather information on the popularity of the German chancellor, she has to collect a sufficiently large sample that is representative of the German population (see Chap. 4 for a discussion of representativeness). She might ask individuals to rate the popularity of the German chancellor on a 0–100 scale. She might also ask respondents about their gender, age, income, education, and place of residency to determine what types of individuals like the head of the German government more and what groups like her

© The Author(s), under exclusive license to Springer Nature Switzerland AG 2023

D. Stockemer and J.-N. Bordeleau, *Quantitative Methods for the Social Sciences*, Springer Texts in Political Science and International Relations,
https://doi.org/10.1007/978-3-031-34583-8_3

less. If these data are collected on a regular basis, it also allows researchers to gain relevant information about trends in societies. For example, so-called trend studies allow researchers to track the popularity of the chancellor over time, and possibly to associate increases and decreases in her popularity with political events such the German reunification in 1990 or the refugee crisis in Germany in 2015.

3.2 A Short History of Survey Research

The origins of survey research go back thousands of years. These origins are linked to the understanding that every society with some sort of bureaucracy, in order to function properly, needs some information about its citizens. For example, in order to set taxation levels and plan infrastructure, bureaucracies need to know basic information about their citizens such as how many citizens live in a geographical unit, how much money they earn, and how many acres of land they own. Hints on first data collection efforts date back to the great civilizations of antiquity, such as China, Egypt, Persia, Greece, or the Roman Empire. A famous example of early data collection is the census mentioned in the bible during the time of Jesus' birth:

> In those days a decree went out from Emperor Augustus that all the world should be registered. This was the first registration and was taken while Quirinius was governor of Syria. All went to their own towns to be registered. Joseph also went from the town of Nazareth in Galilee to Judea, to the city of David called Bethlehem, because he was descended from the house and family of David. He went to be registered with Mary, to whom he was engaged and who was expecting a child. —Luke 2:1–5.

While it is historically unclear whether the census by Emperor Augustus was actually held at the time of Jesus' birth, the citation from the bible nevertheless shows that, as early as in the ancient times, governments tried to retrieve information about their citizens. To do so, families had to register in the birthplace of the head of the family and answer some questions which already resembled our census questions today.

In the middle ages, data collection efforts and surveys became more sophisticated. England took a leading role in this process. The first Norman king, William the Conqueror, was a leading figure in this quest. After his conquest of England in 1066, he strived to gather knowledge on the property conditions, as well as the yearly income of the barons and cities in the seized territories. For example, he wanted to know how many acres of land the barons owned so that he could determine appropriate taxes. In the following centuries, the governing processes became increasingly centralized. To run their country efficiently and to defend the country against external threats, the absolutist English rulers depended on extensive data on labour, military capabilities, and trade (Hooper, 2006). While some of these data were "hard data" collected directly from official books (e.g., the manpower of the army), other data, for example on military capabilities, trade returns or the development of the population, were, at least in part, retrieved through survey questions or interviews. Regardless of its nature, the importance of data collection rose, in particularly, in the economic and military realms. London was the first city where statistics were systematically applied to some collected data. In the seventeenth century, economists including John Graunt, William Petty, and Edmund Halley tried to estimate population developments on the basis of necrologies and birth records. These studies are considered to be the precursors of modern quantitative analysis with the focus on causal explanations (Petty & Graunt, 1899).

Two additional societal developments rendered the necessity for good data the more urgent. First, the adaption of a data-based capitalist economic system in the eighteenth and nineteenth century accelerated data collection efforts in England and later elsewhere in Europe. The rationalization of administrative planning processes in many European countries further increased the need to gain valid data, not only about the citizens, but also about the administrative processes. Again, some of these data could only be collected by asking others. The next boost then occurred in the early nineteenth century. The Industrial Revolution combined with urbanization had created high levels of poverty for many residents in large British cities such as Manchester or Birmingham. To get some 'valid picture' of the diffusion of poverty, journalists collected data by participating in poor people's lives, asking them questions about their living standard and publishing their experiences. This development resulted in the establishment of "Statistical Societies" in most large English cities (Wilcox, 1934). Although, the government shut down most of these statistical societies, it was pressed to extend its own data gathering by introducing routine data collections on births, deaths and crime. Another element of these developments was the implementation of "enquete" commissions whose work looked at these abominable living conditions in some major English cities and whose conclusions were partly based on quantitative data gathered by asking people questions about their lives. Similar developments happened elsewhere, as well. A prominent example is the empirical research of medical doctors in Germany in the nineteenth century, who primarily examined the living and working conditions of laborers and the healthcare system (Schnell et al., 2011: 13–20).

Despite these efforts, it was not until the early twentieth century until political opinion polling in the way we conduct it today was born. Opinion polling in its contemporary form has its roots in the United States of America (USA). It started in the early twentieth century when journalists attempted to forecast the outcomes of presidential elections. Initially, the journalists just took the assessment of some citizens before newspapers came up with more systematic approaches to predict the election results. The "Literary Digest" was the first newspaper to distribute a large number of postal surveys among voters in 1916 (Converse, 2011). The poll also correctly predicted the winner of the 1916 Presidential Elections, Woodrow Wilson. This survey was the first mass survey in the United States and the first systematic opinion poll in the country's history (Burnham et al., 2008: 99 f.). At about the same time, the British philanthropists Charles Booth and Seebohm Rowntree chose interview approaches to explain the causes of poverty. What distinguishes their works from former studies is the close link between social research and political application. To a get valid picture of poverty in the English city of York, Rowntree attempted to interview all working-class people living in York. Of course, this was a long and tiring procedure that took several years. The Rowntree example rendered it very clear to researchers, journalists, and data collection organizations, that collecting data on the full population researchers want to study is very cumbersome and difficult do to do. Consequently, this method of data collection has become very exceptional (Burnham et al., 2008: 100 f.; Schnell et al., 2011: 21–23). Due to the immense costs associated with complete enumerations only governments have the means to carry them out today (e.g., through the census). Researchers must rely mainly on samples, which they use to draw inferences on population statistics. Building on the work of the Literary Digest, in the USA and various efforts on the continent, the twentieth century has seen a refinement of survey and sampling techniques and their broad application to many different scenarios, be they economic, social, or political. Today, surveys are ubiquitous. There is probably not one adult individual in the Western world who has not been asked at least once in her lifetime to participate in a survey.

3.3 The Importance of Survey Research in the Social Sciences and Beyond

Survey research is one of the pillars of social science research in the twenty-first century. Surveys are used to measure almost everything: from voting behavior to public opinion to sexual preferences (Leeuw et al., 2008: 1). Surveys are of interest to wide range of constituents including citizens, parties, civil society organizations and governments. Individuals might be interested in situating their beliefs and behavior in relations to those of their peers and societies. Parties might want to know which party is ahead in the public preference at any given point in time, and what the policy preferences of citizens are. Civil society organizations might use surveys to give credence to their lobbying points. Governments at various levels (i.e., the federal, regional, or local) may use surveys to find out how the public judges their performance or how popular specific policy proposals are among the general public. In short, surveys are ubiquitous in social and political life (for a good description of the importance of survey research see Archer & Berdahl, 2011).

Opinion polls help us to situate ourselves with regard to others in different social settings. On the one hand, survey research allows us to compare our social norms and ideals in Germany, Western Europe, or the Americas to those in Japan, China, or Southeast Asia. For example, analyzing data from a general cross-national social survey provides us with an opportunity to compare attitudes and social behaviors across countries; for instance, we can compare whether we eat more fast food, watch more television, have more pets, or believe more in extensive social welfare than citizens in Australia or Asia. Yet, not only does survey research allow us to detect between country variations in opinions, beliefs and behaviors, but also within a country, if the sample is large enough. In Germany, for example, large-scale surveys can detect if individuals in the East have stronger anti-immigrant attitudes than individuals in the West. In the USA, opinion polls can identify whether the approval rating of former President Trump are higher in Texas than in Connecticut. Finally, opinion polls can serve to detect differences in opinion between different cohorts of the population. For example, we can compare how much trust young people (i.e., individuals in the age cohort 18 to 25) have into the military compared to senior citizens (i.e., individuals aged 60 and older) both for one country and for several countries.

Survey research has also shaped the social- and political sciences as academic disciplines. To illustrate this, we will introduce two classic works in political science whose findings and conclusions are primarily based on survey research. First, one of the most outstanding political treatises based on survey research is *The Civic Culture* by Almond and Verba (1963). In their study, the authors use surveys on political orientations about the political systems (e.g., opinions, attitudes, and values) to detect that cultural norms must be congruent with the political system to ensure the stability of the system in question. Another classic using survey research is Robert Putnam's *Bowling Alone: The collapse and revival of American community* (2001). Mainly through survey research, Putnam finds that social engagement had weakened in the USA during the late twentieth century. He links the drop in all types of social and political activities to a decrease in membership in all kinds of organizations (e.g., social, political, or community organizations), declining contract among individuals (e.g., among neighbors, friends and family), less volunteering, and less religious involvement. It should also be noted that survey research is not only a stand-alone tool to answer many relevant research questions, but it can also be combined with other types of research such as qualitative case studies or the analysis of hard macro-level data. In a prime example of mixed methods, Wood (2003), aiming to understand the peasant's rationales in El Salvador to join revolutionary movements in the country's civil war, uses first ethnographic interviews of some peasants in a specific region to tap into these motivations. In a later stage, she employs large national household surveys to confirm the conclusions derived from the interviews.

3.4 Overview of Some of the Most Widely Used Surveys in the Social Sciences

Governments, governmental-, and non-governmental organizations, and social research centers spend millions of dollars per year to conduct cross national surveys. These surveys (e.g., the World Value Survey, or the European Social Survey) use representative or random samples of individuals in many countries to detect trends in individuals' social and political opinions, as well as their social and political behavior. We can distinguish between different types of surveys. First, behavioral surveys measure individuals political-, health-, or job-related behavior. Probably most prominent in the field of political science, election surveys gauge individuals' conventional and unconventional political activities in a reginal national or international context (e.g., whether somebody participates in elections, engages in protest activity, or contacts a political official). Other behavioral surveys might capture health risk behaviors, employee habits or drug use, just to name few examples. Second, opinion surveys try to capture opinions and beliefs in a society; these questionnaires aim at gauging individual opinions on a variety of topics ranging from consumer behavior to public preferences, to political ideologies, to preferred free time activities and preferred vacation spots.

Below, we present three of the most widely used surveys in political science, and possibly the social sciences more generally: the Comparative Study of Electoral Systems (CSES), the World Value Survey (WVS), and the European Social Survey (ESS). Hundreds if not thousands of articles have emanated from these surveys. In these large-scale research projects, the project managers' duties include the composition of the questionnaire and the selection and training of the interviewers. The latter function as the link between the responsible project managers and respondents. They run the interviews and should record the responses precisely and thoroughly (Loosveldt, 2008: 201).

3.4.1 The Comparative Study of Electoral Systems (CSES)

The Comparative Study of Electoral Systems (CSES) is a collaborative program of cross-national research among election studies conducted in over 50 states. The CSES is composed of three tightly linked parts: first, a common module of public opinion survey questions is included in each participant country's post-election study. These "micro-level" data include vote choice, candidate and party evaluations, current and retrospective economic evaluations, evaluations of the electoral system itself, and standardized socio-demographic measures. Second, district-level data are reported for each respondent, including electoral returns, turnout, and the number of candidates. Finally, the section on system- or "macro-level" data reports aggregate electoral returns, electoral rules and formulas, and regime characteristics. This design allows researchers to conduct cross-level and cross-national analyses, addressing the effects of electoral institutions on citizens' attitudes and behaviors, the presence and nature of social and political cleavages, and the evaluation of democratic institutions across different political regimes.

The CSES is unique among comparative post-electoral studies because of the extent of cross-national collaboration at all stages of the project: the research agenda, the survey instrument, and the study design are developed by the CSES Planning Committee, whose members include leading scholars of electoral politics from around the world. This design is then implemented in each country by that country's foremost social scientists, as part of their national post-election studies. Frequently, the developers of the survey decide upon a theme for any election cycle. For example, the initial round of collaboration focused on three general themes: the impact of electoral institutions on citizens' political cognition and behavior (parliamentary versus presidential systems of government, the electoral rules that govern the casting and counting of ballots; and political parties); the nature of political and social cleavages and alignments; and the evaluation of democratic institutions and processes. The key theoretical question addressed by the second module is the contrast between the view that elections are a mechanism to hold government accountable and the view that they are a means to ensure that citizens' views and interests are properly represented in the democratic process. It is the module's aim to explore how far this contrast and its embodiment in institutional structures influences vote choice and satisfaction with democracy.

The CSES can be accessed at: http://www.isr.umich.edu/cps/project_cses.html

3.4.2 The World Value Survey (WVS)

The World Values Survey is a global research project that explores peoples' values and beliefs, how they change over time, and what social and political impact they have. It emerged in 1981 and was mainly coined by the scientists Ronald Inglehart, Jan Kerkhofs, and Ruud de Moor. The survey's focus was initially on European countries, although since the late 1990, non-European countries have received more attention. Today, more than 80 independent countries representing 85% of the world's population are included in the survey (Hurtienne and Kaufmann 2015: 9f.). The survey is carried out by a worldwide network of social scientists who, since 1981, have conducted representative national surveys in multiple waves in over 80 countries. The WVS measures, monitors, and analyzes a host of issues including support for democracy, tolerance of foreigners and ethnic minorities, support for gender equality, the role of religion and changing levels of religiosity, the impact of globalization, attitudes toward the environment, work, family, politics, national identity, culture, diversity, insecurity, and subjective well-being on the basis of face-to-face interviews. The questionnaires are distributed among 1,100 to 3,500 interviewees per country. The findings are valuable for policy makers seeking to build civil society and democratic institutions in developing countries. The work is also frequently used by governments around the world, scholars, students, journalists and international organizations and institutions such as the World Bank and the United Nations (UNDP and UN-Habitat). Thanks to the increasing number of participating countries and the growing time period that the WVS covers, the WVS satisfies (some of) the demand for cross-sectional attitudinal data. The application of WVS data in hundreds of publications and in more than 20 languages stresses the crucial role that the WVS plays in scientific research today (Hurtienne and Kaufmann 2015: 9f.).

The World Value Survey can be accessed at http://www.worldvaluessurvey.org/

3.4.3 The European Social Survey (ESS)

The European Social Survey (ESS) is an academically driven cross-national survey that has been conducted every two years across Europe since 2001. It is directed by Rory Fitzgerald from the City University of London. The survey measures the attitudes, beliefs, and behavioral patterns of diverse populations in more than thirty European nations. As the largest data collection effort in and on Europe, the ESS has five aims: (1) to chart stability and change in social structure, conditions and attitudes in Europe and to interpret how Europe's social, political, and moral fabric is changing, (2) to achieve and spread higher standards of rigor in cross-national research in the social sciences, including for example, questionnaire design and pre-testing, sampling, data collection, reduction of bias and the reliability of questions, (3) to introduce soundly-based indicators of national progress, based on citizens' perceptions and judgements of key aspects of their societies, (4) to undertake and facilitate the training of European social researchers in comparative quantitative measurement and analysis, and (5) to improve the visibility and outreach of data on social change among academics, policy makers, and the wider public.

The findings of the ESS are based on face-to-face interviews and the questionnaire is comprised of three sections, a core module, two rotating modules and a supplementary questionnaire. The core module comprises questions on the media and social trust, politics, the subjective well-being of individuals, gender and household dynamics, socio-demographics, and social values. As such, the core module should capture topics that are of enduring interest for researchers as well as provide the most comprehensive set of socio-structural variables in a cross-national survey worldwide. The two rotating modules capture "hot" social science topics; for example, rotating modules in 2002 and 2014 focused on immigration, while the 2016 wave captures European citizens' attitudes about welfare and opinions toward climate change. The purpose of the supplementary questionnaire at the end of the survey is to elaborate in more detail on human values and to test the reliability and validity of the items in the principal questionnaire on the basis of some advanced statistical techniques.

The European Social Survey can be accessed at http://www.europeansocialsurvey.org/

3.5 Different Types of Surveys

For political science students, it is important to realize that one survey design does not necessarily resemble another survey design. Rather, in survey research, we generally distinguish between two types of surveys: cross-sectional surveys and longitudinal surveys (see Frees, 2004: 2).

3.5.1 Cross-Sectional Survey

A cross-sectional survey is a survey that is used to gather information about individuals at a single point in time. The survey is conducted once and not repeated. An example of a cross-sectional survey would be a poll that asks respondents in the USA how much they donated toward the reconstruction efforts after Hurricane

Katrina hit the Southern States of the USA. Surveys, such as the one capturing donation patters in the aftermath of Hurricane Katrina, are particularly interesting to seize attitudes and behaviors related to one event that probably will not repeat itself. Yet, cross-sectional surveys are not only used to capture one-time events. To the contrary, they are quite frequently used by researchers to tap into all types of research questions. Because it is logistically complicated, time-consuming and costly to conduct the same study at regular intervals with or without the same individuals, cross-sectional studies are frequently the fall-back option for many researchers. In many instances, the use of cross-sectional surveys can be justified from a theoretical perspective; frequently, a cross-sectional study still allows researchers to draw inferences about relationships between independent and dependent variables (Behnke et al., 2006: 70 f.).

However, it is worth noting that the use of these types of surveys to detect empirical relationships can be tricky. Most importantly, because we only have data at one point for both independent and dependent variable, cross-sectional surveys cannot establish causality (i.e., they cannot establish that a change in the independent variable precedes a change in the dependent variable) (De Vaus, 2001: 51). Therefore, it is important that findings/conclusions derived from cross-sectional studies are supported by theory, logic, and/or intuition (Frees, 2004: 286). In other words, a researcher should only use cross-sectional data to test theories, if the temporal chain between independent and dependent variable is rather clear a priori.

If we have clear theoretical assumptions about a relationship, a cross-sectional survey can provide a good tool to test hypotheses. For example, a cross-sectional survey could be appropriate to test the linkage between formal education and casting a ballot at elections, as there is a strong theoretical argument in favor of the proposition that higher formal education will increase somebody's propensity to vote. According to the resource model of voting (see Brady et al., 1995), higher educated individuals have the material and non-material resources necessary to understand complex political scenarios, as well as the network connections; all of which should render somebody more likely to vote. Vice versa, uneducated individuals lack these resources and are frequently politically disenfranchised. Practically, it is also impossible that the sheer act of voting changes somebody's formal education. Hence, if we analyze a cross-sectional survey on voting and find that more educated individuals are more likely to vote, we can assume that this association reflects an empirical reality. To take another example, if we want to study the influence of age on protesting, data from a cross-sectional survey could be completely appropriate, as well, to study this relationship, as the causal change clearly goes from age to protesting and not the other way round. By definition, the fact that an individual protests does not make them younger or older, at least when we look at somebody's biological age.

Nevertheless, empirical relationships are not always that clear-cut. Rather contrary, sometimes it is tricky to derive causal explanations from cross-sectional studies. To highlight this dilemma let us take an example from American Politics and look at the relationship between watching Fox news and voting for Donald Trump. For one, it makes theoretical sense that watching *Fox News* in the United

States increases somebody's likelihood to vote for Donald Trump in the Presidential Election, because this TV chain supports this populist leader. Yet, the causal or correlational chain could also go the other way round. In other words, it might also be that somebody, who decided to vote for Trump is actively looking for a news outlet that follows her convictions. As a result, she might watch *Fox News* after voting for Trump.

A slightly different example highlights even clearer that the correlational or causal direction between independent and dependent variable is not always clear. For example, take the following example, it is theoretically unclear if the consumption of *Fox News* renders somebody more conservative, or if more conservative individuals have a higher likelihood to watch Fox News. Rather than one variable influencing the other, both factors might mutually reinforce each other. Therefore, even if a researcher finds support for the hypothesis that watching Fox News makes people more conservative, we cannot be sure of the direction of this association because a cross-sectional survey would ask individuals the same question at the same time.[1] Consequently, we cannot detect what comes first; watching Fox News or being conservative. Hence, cross-sectional surveys cannot resolve the aforementioned temporal aspect. Rather than a cross-sectional survey, a longitudinal survey would be necessary to determine the causal chain between being conservative and watching *Fox News*. Such a survey, in particular, if it is conducted over many years and if it solicits the same individuals in regular intervals, could tell researchers if respondents first become conservative and then watch *Fox News*, or if the relationship is the other way round.

3.5.2 Longitudinal Survey

In contrast to cross-sectional studies, which are conducted once, longitudinal surveys repeat the same survey questions several times. This allows the researchers to analyze changing attitudes or behaviors that occur within the population over time. There are three types of longitudinal surveys: trend studies, cohort studies, and panel studies.

Trend Surveys

A Trend Study, which is frequently also labeled a repeated cross-sectional survey, is a repeated survey that is normally not composed of the same individuals in the different waves. Most of the main international surveys including the *European Social Survey* or the *World Value Survey* are trend studies. The surveys of the different waves are fully or partly comprised of the same questions. As such they allow researchers to detect broad changes in opinions and behaviors over time. Nevertheless, and because the collected data covers different individuals in each wave of the study, the collected data merely allows for conclusions at the

[1] In the literature, such reversed causation is often referred to as an endogeneity problem.

aggregate level such as the regional or the national level (Schumann, 2012: 113). To highlight, most of the major surveys ask the question: How satisfied are you with how democracy works in your country? Frequently, the answer choices range from 0 or not satisfied at all to 10 or very satisfied. Since citizens answer these questions every two years, researchers can track satisfaction rates with democracy over a longer period such as 10 years. Comparing the answers for several waves, a researcher can also establish if the same or different independent variables (e.g., unemployment or economic insecurity, gender, age, or income) trigger higher rates of dissatisfaction with democracy. However, what such a question/ study cannot do is to track down what altered an individual's assessment of the state of democracy in her country. Rather, it only allows researchers to draw conclusions at the macro- or aggregate level.

Cohort Surveys

While trend studies normally focus on the whole population, cohort studies merely focus on a segment of the population. One common feature of a cohort study is that a central event or feature occurred approximately at the same time to all members of the group. Most common are birth cohorts. In that case, birth is the special event that took place in the same year or in the same years for all members of the cohort (e.g., all Americans who were born in or after 1960). Analogous to trend Studies, cohort studies use the same questions in several waves. In each wave, a sample is drawn from the cohort. This implies that the population remains the same over time, whereas the individuals in the sample change. A typical example of cohort studies is the "British National Child Study" (NCDS). In the course of this study, 11,400 British citizens born between March 3rd and 9th 1958 were examined with respect to their health, education, income, and attitudes in eight waves in a time span of 50 years (Schnell et al., 2011: 237 f.).

Panel Surveys

Panel Studies normally ask the same questions to the same people in subsequent waves. These types of surveys are the costliest and most difficult to implement, but they are the best suited to detect causal relationships or changes in individual behavior. For example, a researcher could ask questions on the consumption of Fox News and being conservative to the same individual over the period of several years. This could help her detect the temporal chain in the relationship between a certain type of news consumption and political ideologies. Panel studies frequently have the problem of high attrition or mortality rates. In other words, people drop out during waves for multiple reasons, for example they could move, become sick or simply refuse further participation. Hence, it is likely that a sample that was representative from the outset becomes less and less representative for subsequent waves of the panel. To highlight, imagine that a researcher is conducting a panel on citizens' preference on which electoral system should be used in a country, and they ask this question every two years to the same individuals. Individuals who are interested in electoral politics, and/or have a strong opinion in favor of one or the other type, might have a higher likelihood to stay in the sample than citizens, you do not care. In contrast, those who are less interested will no longer participate in future waves. Others, like old age citizens might die or move into an old peoples' home. A third group such as diplomats and consultants is more likely to move than manual workers. It is possible to continue the list. Therefore, there is the danger that many panels become less representative of the population for any of the waves covered. Nevertheless, in particular, if some representativeness remains in subsequent waves or if the representativeness is not an issue for the research question, panel studies can be a powerful tool to detect causal relationships. An early example of an influential panel study is Butler and Stokes' *Political Change in Britain: Forces Shaping Electoral Choice.* Focusing on political class as the key independent variable for the vote choice for a party, the authors conducted three waves of panels with the same randomly selected electors in the summer 1963, after the general elections 1964 and after the general elections 1966 to determine habitual voting and vote switching. Among others, they find that voting patterns in favor of the three main parties (i.e., the Liberal Party, Labour, and the Conservative Party) are more complex to be fully captured by class.

References

Almond, G., & Verba, S. (1963). *[1989]: The civic culture: Political attitudes and democracy in five nations.* Sage Publications.

Archer, K., Berdahl L. (2011). *Explorations: Conducting empirical research in Canadian political science.* Oxford: University Press.

Behnke, J., & Baur, N./Behnke, N. (2006). *Empirische Methoden der Politikwissenschaft.* Schöningh.

Brady, H. E., Verba, S., & Schlozman, K. L. (1995). Beyond SES: A resource model of political participation. *American Political Science Review, 89*(2), 271–294.

Burnham, P., Lutz, G. L., Grant, W., & Layton-Henry, Z. (2008). *Research methods in politics* (2nd ed.). Palgrave Macmillan.

Converse, J. M. (2011). *Survey research in the United States: Roots and emergence 1890–1960.* Transaction Publishers.

De Leeuw, E. D., Hox, J. J., & Dillman D. A. (2008). *The cornerstones of survey research.* In: E. D. De Leeuw, J. J. Hox, & D. A. Dillman (Eds.). *International handbook of survey methodology.* Lawrence Erlbaum Associates.

De Vaus, D. (2001). *Research design in social research.* Sage Publications.

ESS (European Social Survey). (2017). *Source questionnaire.* http://www.europeansocialsurvey.org/methodology/ess_methodology/source_questionnaire/. August 07, 2017.

Fowler, F. J. (2009). *Survey research methods* (4th ed.). Sage Publications.

Frees, E. W. (2004). *Longitudinal and panel data: Analysis and applications in the social sciences.* Cambridge University Press.

Hooper, K. (2006). Using William the Conqueror's accounting record to assess manorial efficiency: A critical appraisal. *Accounting History, 11*(1), 63–72.

Hurtienne, T., & Kaufmann, G. (2015). *Methodological biases: Inglehart's world value survey and Q methodology.*

Loosveldt, G. (2008). *Face-to-face interviews.* In: E. D.De Leeuw, J. J. Hox, & D. A. Dillman (Eds.), *International Handbook of survey methodology.* Lawrence Erlbaum Associates.

Petty, W., & Graunt, J. (1899). *The economic writings of Sir William Petty* (Vol. 1). University Press.

Putnam, R. D. (2001). *Bowling alone: The collapse and revival of American community*. Simon and Schuster.
Schnell, R., Hill, P. B., & Esser, E. (2011). *Methoden der empirischen Sozialforschung* (9th ed.). Oldenbourg.
Schumann, S. (2012). *Repräsentative Umfrage: Praxisorientierte Einführung in empirische Methoden und statistische Analyseverfahren* (6th ed.). Oldenbourg.
Willcox, W. F. (1934). Note on the chronology of statistical societies. *Journal of the American Statistical Association, 29*(188), 418–420.
Wood, E. J. (2003). *Insurgent collective action and civil war in El Salvador*. Cambridge University Press.

Further Reading

Why do we need survey research?

Converse, J. M. (2017). *Survey research in the United States: Roots and emergence 1890–1960*. Routledge.

This book has more of an historical ankle. It tackles the history of survey research in the USA.

Davidov, E., Schmidt, P., & Schwartz, S. H. (2008). Bringing values back in: The adequacy of the European Social Survey to measure values in 20 countries. *Public Opinion Quarterly, 72*(3), 420–445.

This rather short article highlights the importance of conducting a large pan-European survey to measure European's social and political beliefs.

Schmitt, H., Hobolt, S. B., Popa, S. A., & Teperoglou, E. (2015). European parliament election study 2014, voter study. *GESIS Data Archive, Cologne. ZA5160 Data file Version, 2*(0).

The European Voter Study is another important election study that researchers and students can access freely. It provides a comprehensive battery of variables about voting, political preferences, vote choice, demographics, and political and social opinions of the electorate.

Applied Survey Research

Almond, G. A., & Verba, S. (1963). *The civic culture: Political attitudes and democracy in five nations*. Princeton University Press.

Almond's and Verba's masterpiece is a seminal work in survey research measuring citizens' political and civic attitudes in key Western democracies. The book is also of the first books that systematically uses survey research to measure political traits.

Inglehart, R., & Welzel, C. (2005). *Modernization, cultural change, and democracy: The human development sequence*. Cambridge University Press.

This is an influential book, which uses data from the World Value Survey to explain modernization as a process that changes individuals' values away from traditional and patriarchal values and toward post-materialist values including environmental protection, minority rights, and gender equality.

4 Constructing a Survey

4.1 Types of Questions a Researcher Can Ask

In principle, in a survey, a researcher can ask questions about what people think, what they do, what attributes they have, and how much knowledge they have about an issue.

Questions about opinions, attitudes, beliefs, values—capture what people think about an issue, a person or an event. These questions frequently give respondents some choices in their answers.

> *Example*: Do you agree with the following statement: The European Union should create a crisis fund in order to be able to rapidly bail out member countries, when they are in financial difficulties. (Possible answer choices: Agree, partly agree, partly disagree, and do not agree)

Questions about individual behavior—capture what people do.

> *Example*: Did you vote in the last election? (Possible answer choices, yes or no)

Questions about attributes—what are peoples' characteristics.

> *Example*: How old are you? (Possible answers are numerical values)
> *Example*: What is your gender? (Possible answers, male, female, other)

Questions about knowledge—how much people know about political, social or cultural issues.

© The Author(s), under exclusive license to Springer Nature Switzerland AG 2023

D. Stockemer and J.-N. Bordeleau, *Quantitative Methods for the Social Sciences*, Springer Texts in Political Science and International Relations,
https://doi.org/10.1007/978-3-031-34583-8_4

Example: In your opinion, how much of its annual GDP per capita does Germany attribute toward development aid?

4.2 Ordering of Questions

There are some general guidelines for constructing surveys that make it easy for participants to respond.

1. The ordering of questions should be logical to the respondents and flow smoothly from one question to the next. As a general rule, questions should go from general to specific, impersonal to personal, and easy to difficult.
2. Questions related to the same issue should be grouped together. For example, if a survey captures different forms of political participation, questions capturing voting, partaking in demonstrations, boycotting and signing petitions should be grouped together. Ideally, the questions could also go from conventional political participation to unconventional political participation. The same applies to other sorts of common issues such opinions about related subjects (e.g., satisfaction with the government's economic, social, and foreign policy, respectively). Consequently, it makes sense to divide the questionnaire in different sections, each of which should begin with a short introductory sentence. However, while it makes sense to cluster questions by theme, there should not be too many similar questions, with similar measurements either. In particular, the so-called consistency bias might come into play, if there are too many similar questions. Cognitively, some respondents wish to appear consistent in the way they answer the questions, in particular if these questions look alike. To mitigate the consistency bias effects, it is useful to switch up the questions and do not have a row of 20 or 30 questions that are openly interrelated (Weisberg et al., 1996: 89f.).

4.3 Number of Questions

Researchers must always draw a fine line between the exhaustiveness and the parsimony of the questionnaire. They want to ask as many questions as are necessary to capture the dependent and all necessary independent variables in the research project. Yet, they do not want to ask too many questions either, because the more questions that are included in an interview or a survey, the less likely people are to finish a face-to-face or phone interview or to return a completed paper or online questionnaire. There are no strict provisions for the length of a survey, though. The general rule that guides the construction of a social science theory could also guide the construction of a questionnaire; a questionnaire should include as many questions as necessary, and as few questions as possible.

4.4 Getting the Questions Right

When writing surveys, researchers normally follow some simple rules. The question wording needs to be clear, simple, and precise. In other words, questions need to be written so that respondents understand their meaning right away. In contrast, poorly written questions lead to ambiguity and misunderstandings and can lead to untrustworthy answers. In particular, vague questions, biased/value laden questions, threatening questions, complex questions, negative questions, and pointless questions should be avoided.

Vague questions:

Vague questions are questions that do not clearly communicate to the respondent what the question is actually all about. For example, a question like "Taken altogether, how happy are you with German Chancellor Scholz?" is unclear. It is imprecise because the respondent does not know what substantive area the questions is based on. Does the pollster want to know whether the respondent is "happy" with his rhetorical style, his appearance, his leadership style, or his government's record, or all of the above? Also, the word happy should be avoided because it is at least somewhat value laden. Hence, the researcher should refine her research question by specifying a policy area, a measurement scale and by using more neutral less colloquial language. Hence, a better question would be: Overall, how would you rate the performance of Chancellor Scholz in the domain of environmental politics in 2022/2023 on a 0 to 100 scale?

Biased or value-laden questions:

Biased or value laden questions are questions that predispose individuals to answer a question in a certain way. These questions are not formulated in a neutral way. Rather, they use strong normative words. Consider, for example, the question: On a scale from 0 to 100, how evil do you think that the German Christian Democratic Party (CDU) is? This question is clearly inappropriate for a survey as it bears judgement on the question's subject. Therefore, a better formulation of the same question would be: On scale from 1 to 100 how would you rate the performance of the Christian Democratic Party in the past year? Ideally, the pollster could also add a policy area to this question.

Threatening questions:

Threating questions might render respondents to surveys uneasy, and/or make it hard for the respondent to answer to her best capacities. For instance, the question: "Do you have enough knowledge about German politics to recall the political program of the four parties the Christian Democrats, the Social Democrats, the Green Party, and the Party of Democratic Socialism?" might create several forms of uneasiness on the beholder. The person surveyed might question their capability to answer the question, they might assume that the pollster is judging them, and they might not know how to answer the question, because it is completely unclear what enough knowledge means. Also, it might be better to reduce the question to one party, because citizens might know a lot about one party program and relatively few things about another program. A better question would be: On a scale from 0 to 100, how familiar are you with the political programs of the Social Democratic Party for the General Election in 2021. (0 means I am not familiar at all and 100 means I am an expert).

Complex questions:

Researcher should avoid complex questions that ask the polled about various issues at once. For example, a question like this: On a scale from 1 to 10, please rate for each of the 12 categories listed below, your level of knowledge, confidence, and experience should be avoided, as it confuses respondents and makes it impossible for the respondent to answer precisely. Rather than clustering many items at once, the research should ask one question per topic.

Negative questions:

The usage of negative questions might induce bias or normative connotation in a question. A question such as "On a scale from 0 to 100 how unfit do you think American President Donald Trump is for Presidency of the United States?" is more value laden then the same question expressed in positive terms: "On a scale from 0 to 100 how fit do you think American President Donald Trump is for the Presidency of the United States?" Using the negative form might also confuse people. Therefore, it is a rule in survey research to avoid the usage of negative wording. This includes the use of words such as "not", "rarely", "never", or words with negative prefixes "in-", "im-", "un-". As a rule, researchers should always ask their questions in a positive fashion.

Pointless questions:

A pointless question is a question that does not allow the researcher or pollster to gain any relevant information. For example, the I-94 form the US Citizenship and Immigration Services that every foreigner that enters the USA must fill out does make any sense. The form asks all foreigner entering the USA the following: Have you ever been or are you now involved in espionage or sabotage, or in terrorist activities or genocide or between 1933 and 1945 were you involved, in any way, in persecutions associated with Nazi Germany or its allies? Respondents can circle either yes or no. This question is futile in two ways. First, people involved in any of these illegal activities have no incentive to admit to it; admitting to it would automatically mean that their entry into the US would be denied. Second, and possibly even more importantly, there remain very few individuals alive who, in theory, could have been involved in illegal activities during the Nazi era. And furthermore, virtually all of those very few individuals who are still alive are probably too old now to travel to the USA.

4.5 Social Desirability

The "social desirability paradigm" refers to a phenomenon that we frequently encounter in social science research: survey respondents are inclined to give socially desirable responses, or responses that make them look good against the background of social norms or common values. The social desirability bias is most prevalent in the field of sensitive self-evaluation questions. Respondents can be inclined to avoid disapproval, embarrassment or legal consequences, and therefore opt for an incorrect answer in a survey. For example, respondents know that voting in elections is a socially desirable act. Hence, they might be tempted to affirm that they have voted in the last presidential or parliamentary election even if they did not. Vice versa, the consumption of hard drugs such as cocaine or ecstasy are socially undesirable acts that are also punishable by law. Consequently, even if the survey is anonymously conducted, a respondent might not admit to consuming hard drugs, even if she does so regularly.

The social desirability construct jeopardizes the validity of a survey as socially undesired responses are underestimated. In a simple survey, it is very difficult for a researcher to detect how much the social desirability bias impacts the given responses. One implicit way of detection would be to check if a respondent answers a number of similar questions in a socially desirable manner. For example, if a researcher asks if respondents have ever speeded, smoked cigarettes, smoked marijuana, cheated with the taxes, or lied to a police officer, and gets negative answers for all 5 items, there is a possibility that the respondent chose social desirability over honesty. Yet, the researcher a priori cannot detect whether this cheating with the answers happened and for what questions. The only thing she can do is to treat the answers of this particular questionnaire with care, when analyzing the questionnaire. Another, albeit also imperfect way, to detect socially desirable rather than correct answers is through follow up questions. For example, a researcher could first ask the question. Did you vote in the last parliamentary election? A follow-up question could then ask the respondent for which party she voted for? This follow up question could include the do not know category. It is likely that respondents remember for which party they voted, in particularly, if the survey is conducted relatively shortly after the election. This implies that somebody, who indicated that she voted, but does not recall her party choice, might be a candidate for a socially desirable rather than an honest answer for the voting question (see Hoffmann, 2014: 7 f.; Steenkamp et al., 2010: 200–202; Van de Vijver & He, 2014: 7 for an extensive discussion of the social desirability paradigm).

Not only can social desirability be inherent in attitudinal and behavioral questions, but it can also come from outside influences such as from the pollster itself. A famous example of the effect of socially desirable responding was the 1989 Virginia gubernatorial race between the African American Douglas Wilder and the Caucasian Marshall Coleman. Wilder, who had been ahead in the polls by a comfortable buffer of 4–11%, ended up winning the election merely by the tiny margin of 0.6%. The common explanation for that mismatch between the projected vote margin in polls and the real vote margin was that a number of white respondents interviewed by African Americans declared their alleged preference for Wilder in order to appear tolerant in the eyes of the African American interviewer (Krosnick, 1999: 44 f.)

4.6 Open-Ended and Closed-Ended Questions

In survey research, we normally distinguish between two broad types of questions: open-ended and closed-ended questions. Questionnaires can be open-ended, closed-ended, or can include a mixture of open- and closed-ended questions. The main difference between these types of questions is that open-ended questions allow respondents to come up with their own answers in their own words, while closed-ended questions require them to select an answer from a set of predetermined choices (see Table 4.1, for a juxtaposition between open- and closed-ended questions). An example of an open-ended question would be: "Why did you join the German Christian Democratic Party?" To answer such a question, the person surveyed must describe, in her own words, what factors enticed her to become a member of that specific party and what her thought process was before joining. In contrast, an example of a closed-ended question would be: "Did you partake in a demonstration in the past 12 months?" This question gives respondents two choices—they can either choose the affirmative answer or the negative answer. Of course, not all open-ended questions are qualitative in nature. For example, the question, "how much money did you donate to the Christian Democratic Party's election campaign in 2012" requires a precise number. In fact, all closed- and open-ended questions either require numerical answers, or answers that can relatively easily be converted into a number so that they can be utilized in quantitative research (see Table 4.1).

In fact, the usage of non-numerical open-ended questions and closed-ended questions frequently follows different logics. Open-ended questions are frequently used in more in-depth questionnaires and interviews aiming to generate high quality data that can help researchers generate hypotheses and/or explain causal mechanisms. Since it can sometimes be tricky to coherently summarize qualitative

Table 4.1 Open-ended versus closed-ended questions

Open-ended questions	Closed-ended questions
No predetermined responses given	Designed to obtain predetermined responses
Respondent is able to answer in his or her own words	Possible answers: yes/no; true false, scales, values
Useful in exploratory research and to generate hypotheses	Useful in hypotheses testing
Require skills on the part of the researcher in asking the right questions	Easy to count, analyze and interpret
Answers can lack uniformity and be difficult to analyze	There is the risk that the given question might not include all possible answers

interview data, qualitative researchers must frequently utilize sophisticated data analytical techniques including refined coding schemes and advanced analytical techniques (Seidman, 2013; Yin, 2015). However, since this is a book about quantitative methods and survey research, we do not discuss qualitative interviewing and coding in detail here. Rather, we focus on quantitative research. Yet, working with a set of categories is not without caveats either, it can lead to biased answers and restrict the respondent in his or her answering possibilities. As a rule, the amount of response choices should not be too restricted to give respondents choices. At the same time, however, the number of options should not exceed the cognitive capacities of the average respondent either. With regards to a phone or face-to-face interview, there is some consensus that not more than seven response categories should be offered (Schumann, 2012: 74).

The choice of the appropriate number of response choices can be a tricky process. As a general rule, it makes sense to be consistent with respect to the selection of response scales for similar questions to prevent confusing the respondent (Weisberg et al., 1996: 98). For example, if a researcher asks in a survey of citizens' satisfaction with the country's economy, the country's government, and the country's police forces, it makes sense to use the same scale for all three questions. However, there are many options we can use. Consider the following question: How satisfied are you with the economic situation in your country? A researcher could use a scale from 0 to 10 or a scale from 0 to 100, both of which would allow respondents to situate themselves. A researcher could also use a graded measure with or without a neutral category and/or a do not know option (for a list of options see Table 4.2). To a certain degree, the chosen operationalization depends on the purpose of the study and the research question, but to a certain degree is also a question of taste.

Table 4.2 Response choices for the question of economic satisfaction

Operationalization 1	Scale from 0 to 100 (zero meaning not satisfied at all, 100 meaning very satisfied)
Operationalization 2	Scale from 0 to 10 (zero meaning not satisfied at all, 10 meaning very satisfied)
Operationalization 3	5 value scale including the categories, very satisfied, satisfied, neutral, not satisfied, not satisfied at all
Operationalization 4	4 value scale including the categories, very satisfied, satisfied, not satisfied, not satisfied at all
Operationalization 5	6 value scale including the categories, very satisfied, satisfied, neutral, not satisfied, not satisfied at all, and I do not know
Operationalization 6	5 value scale including the categories, very satisfied, satisfied, not satisfied, not satisfied at all, I do not know

4.7 Types of Closed-Ended Survey Questions

Researchers can choose from a variety of questions including scales, dichotomous questions and multiple-choice questions. In this section, we present the most important types of questions:

4.7.1 Scales

Scales are ubiquitous in social sciences questionnaires. Nearly all opinion-based questions use some sort of scale. An example would be: "Rate your satisfaction with your country's police forces from 0 (not satisfied at all) to 10 (very satisfied)". Many questions tapping into personal attributes use scales, as well (e.g., "how would you rate your personal health?" (0 = poor; 1 = rather poor, 2 average, 3 rather good, 4 very good)). Finally, some behavioral questions also use scales. For example, you might find the question: "How often do you generally watch the news on TV?" Response options could be not at all (coded 0), rather infrequently (coded 1), sometimes (coded 2), rather frequently (coded 3), and very frequently (coded 4). The most prominent scales are Likert and Guttman scales.

Likert Scale: A Likert scale is the most frequently used ordinal variable in questionnaires (maybe the most frequently used type of question overall) (Kumar, 1999: 129). Likert scales use fixed choice response formats and are designed to measure attitudes or opinions (Bowling, 1997; Burns & Grove, 1997). Such a scale assumes that the strength/intensity of the experience is linear, i.e., on a continuum from strongly agree to strongly disagree, and assumes that attitudes can be measured. Respondents may be offered a choice of five to seven or even nine pre-coded responses with the neutral point being neither agree nor disagree (Kumar, 1999: 132).

Example 1 Going to war in Iraq was a mistake of the Bush administration.

The respondent has 5 choices: strongly agree, somewhat agree, neither agree nor disagree, somewhat disagree, strongly disagree.

Example 2 How often do you discuss politics with your peers?

The respondent has 5 choices: very frequently, frequently, occasionally, rarely, never.

In the academic literature, there is a rather intense debate about the usage of the neutral category and the do not know option. This applies to Likert scales and other survey questions as well. Including a neutral category makes it easier for some people to choose. It might be a safe option, in particular for individuals, who prefer not taking a choice. However, excluding the neutral option forces individuals to take a position, even if they are torn, or think that the economy functions neither good nor badly for the precise question at hand (Neuman & Robson, 2014; Snapford, 2006). A similar logic applies to the do not know option (see Mondak & Davis, 2001; Sturgis et al., 2008). The disadvantage of this option is that many respondents go with "no opinion" just because it is more comfortable. Yet, in other cases, respondents choose this option due to conflicting views and not due to a lack of knowledge. So, on the one hand, there is an argument to be made that citizens usually do have an opinion and lean at least slightly to one side or another. That is why, the inclusion of a "no opinion" option can reduce the number of respondents who give answers to more complex subjects. On the other hand, if this option is omitted respondents can feel pressed to pick a side although they really are indifferent for example because they know very little about the matter. Pushing respondents to give arbitrary answers can affect the validity and reliability of a survey.

For example, somebody might not be interested in economics and besides getting her monthly paycheck has no idea how the economy functions in her country. Such a person can only take a random guess for this question if the "do not know" option is not included in the questionnaire. For sure, she could leave out the question, but this might feel challenging. Hence, the respondent might feel compelled to give an answer even if it is just a random guess.

While there is no panacea for avoiding these problems, the researcher must be aware of the difficulty of finding the right number and type of response categories and might think through these categories carefully. A possible way of dealing with this challenge is to generally offer "no opinion" options, but to omit them for items on well-known issues or questions where individuals should have an opinion. For example, if you ask respondents about their self-reported health, there is no reason to assume that respondents do not know how they feel. In contrast, if you ask them knowledge questions (e.g., "who is the president of the European Commission?") the do not know option is a viable choice, as it is highly possible that politically uninterested citizens do not know, who the president of the European Commission is. Alternatively, the researcher can decide to omit the "no opinion" option and test the strength of an attitude through follow-up questions. This way, it might become evident whether the given response represents a real choice rather than a guess (Krosnick, 1999: 43 f.; Weisberg et al., 1996: 89 f.). For example, a pollster could first ask an individual about her political knowledge using the categories low, rather low, middle, rather high and very high. In a second step, she could ask "hard" questions about the function of the political system or the positioning of political parties. These "hard" questions could verify whether the respondent judges her political knowledge level accurately.

The Semantic Differential Scale: The use of the Semantic Differential Scale allows for more options than the use of a Likert scale, which is restricted to four or five categories. Rather than having each category labeled, the semantic scale uses two bipolar adjectives at each end of the question. Semantic differential scales normally range between 7 and 10 response choice (see Tables 4.3. and 4.4).

Table 4.3 Example of a semantic differential scale with 7 response choices

How satisfied are you with the services received?						
1						7
Not at all satisfied						Very satisfied

Table 4.4 Example of a semantic differential scale with 10 response choices

Do you think immigrants make your country a better or a worse place to live in?									
1									10
Better place									Worse place

Large Rating Scales: These are scales that have a larger range than 0 to 10. Such scales can often have a range from 0 to 100. Within this range, the respondent is free to choose the number that most accurately reflects her opinion. For examples, researchers could ask respondents, how satisfied they are with the state of the economy in their country, on a scale from 0 not satisfied at all, to 100 very satisfied.

Guttman Scale: The Guttman scale represents another rather simple way of measuring ordinal variables. This scale is based on a set of items with which the respondents can agree or disagree. All items refer to the exact same variable. However, they vary in their level of "intensity" which means that even people with low agreement might still agree with the first items (questions with a low degree of intensity usually come first) while it takes high stakes for respondents to agree with the last ones. The respondent's score is the total number of items he agrees with—a high score implies a high degree of agreement with the initial questions. Since, all items of one variable are commonly listed in increasing order according to their intensity, this operationalization assumes that if a respondent agrees with any one of the asked items, she should have agreed with all previous items too. The following example clarifies the idea of "intensity" in response patterns.

***Example*: Do you agree or disagree that abortions should be permitted**

1. When the life of the woman is in danger
2. In the case of incest or rape
3. When the fetus appears to be unhealthy
4. When the father does not want to have a baby
5. When the woman cannot afford to have a baby
6. Whenever the woman wants.

Due to the Guttman scale's basic assumption, a respondent with a score of 4 agrees with the first four items and disagrees with the last two items. The largest challenge when using a Guttman scale is to find a suitable set of items that (perfectly) match the required response patterns. In other words, there must be graduation between each response category and consensus that agreement to the fourth item is more difficult and restricted than agreement to the third item.

4.7.2 Dichotomous Survey Question

The dichotomous survey question normally leaves respondents with two answering choices. These two choices can include personal characteristics such as male and female, or they can include questions about personal attributes such as whether somebody is married or not.

> *Example*: Did you participate in a lawful demonstration in the past 12 months? (yes/no)

Please note that some questions that were asked dichotomously for decades have recently been asked in a more trichotomous fashion. The prime example is gender. For decades, survey respondents had two options to designate their gender as man or woman, not considering that some individuals might not identify with either of the dichotomous choices. Thus, the politically correct way, now, is to include a third option such as neither/other.

4.7.3 Multiple-Choice Questions

Multiple-choice questions are another form of a frequently used question type. They are easy to use, respond to, and analyze. The different categories the respondents can choose from must be mutually exclusive. Multiple-choice questions can be both single answer (i.e., the respondent can only pick one option) or multiple answer (i.e., the respondent can pick several answers) (see Tables 4.5 and 4.6).

One of the weaknesses of some type of multiple-choice questions is that the choice of responses is rather limited and sometimes not final. For example, in Table 4.6, somebody could receive her political information from weekly magazines, talk shows, or political parties. In the current example, we have not included these options for parsimony reasons. Yet, the option E (other) allows respondents to add other items, thus allowing for greater inclusivity (But please note, that if

Table 4.5 Example of single answer multiple-choice question

What is your favorite politician among the 4 politicians listed below?		
A	Donald Trump	
B	Emmanuel Macron	
C	Angela Merkel	
D	Theresa May	

Table 4.6 Example of a multiple answer multiple-choice question

Which medium do you use to get your political information from? (click all that apply)		
A	Television news broadcasts	
B	Radio	
C	Internet	
D	Newspaper	
E	Discussion with friends	
F	Other (please specify)	
G	I do not inform myself politically	

the respondents list too many options under the category "other", the analysis of the survey becomes more complicated.).

4.7.4 Numerical Continuous Questions

Numerical continuous questions ask respondents a question that in principle allows the respondent to choose from an infinite number of response choices. Examples would be questions that ask: what is your age? What is your net income?

4.7.5 Categorical Survey Questions

Frequently, researchers split continuous numerical questions into categories. These categories frequently correspond to established categories in the literature. For example, instead of asking what your age is, you could provide age brackets distinguishing young individuals (i.e., 18 and younger), rather young individuals (19–34), middle aged individuals (35–50), rather old individuals (51–64) and old individuals (65 and higher) (see Table 4.7.). The same might apply to income. For example, we could have income brackets for every $ 20,000 (see Table 4.8).

Using brackets or categories might be particularly helpful for features, where respondents do not want to reveal the exact number. For example, some respondents might not want to reveal their exact age. The same might apply to income.

Table 4.7 Example of a categorical survey question with 5 choices

Which category best describes your age?	
	Younger than 18
	19–34
	35–50
	51–64
	65 and higher

Table 4.8 Example of a categorical survey question with 6 choices

In which bracket does your net income fall?	
	Lower than $20,000
	$20,000 to $39,999
	$40,000 to $59,999
	$60,000 to $79,999
	$80,000 to $99,999
	Over $100,000

Table 4.9 Example of a rank order question

Rank the following five parties from most popular (coded 1) to least popular (coded 5)	
	Christian Democratic Party
	Social Democratic Party
	Free Democratic Party
	Green Party
	Left Party
	Alternative for Germany

If they are offered a rather broad age or income bracket, they might be more willing to provide an answer compared to when we probe them for their exact age or income.

4.7.6 Rank Order Questions

Sometimes researchers might be interested in rankings. Rank-order questions allow respondents to rank persons, brands, or products based on certain attributes such as the popularity of politicians or the vote choice for parties. In rank order questions, the respondent must consecutively rank the choices from most favorite to least favorite (see Table 4.9).

4.7.7 Matrix Table Questions

Matrix level questions are questions in tabular format. They consist of multiple questions with the same response choices. The questions are normally connected to each other, and the response choices frequently follow a scale such as a Likert scale (see Table 4.10).

Table 4.10 Example of matrix table question

How would you rate the following attributes of President Trump?					
	Well below average	Below average	Average	Above Average	Well above average
Honesty					
Competence					
Trustworthiness					
Charisma					

4.8 Different Variables

Regardless of the type of survey questions, there are four different ways to operationalize survey questions. The four types of variables are: string variables, continuous variables (interval variables), ordinal variables, and nominal variables.

A **string variable** is a variable that normally occupies the first column in a dataset. It is a non-numerical variable, which serves as the identifier. Examples are individuals in a study, or countries. This variable is normally not part of any analysis.

A **continuous variable** can have, in theory, an infinite number of observations (e.g., age, income). Such a question normally follows from a continuous numerical question. To highlight, personal income levels can in theory have any value between 0 and infinity.

An **interval variable** is a specific type of variable; it is a continuous variable with equal gaps between values. For example, counting the income in steps of thousand would be an interval variable: 1000, 2000, 3000, 4000 ….

A **nominal variable** is a variable with several categories. However, there is no specific order or value to the categorization (e.g., the order is arbitrary). Because there is no hierarchy in the organization of the data, this type of variable is the most difficult to present in datasets (see discussion under Sect. 4.9.1.).

- An example of a two-categorical nominal variable would be gender (i.e., men and women). Such a variable is also called **dichotomous**, or **dummy** variable.
- An example of a categorical variable with more than two categories would be religious affiliation (e.g., Protestant, Catholic, Buddhist, Muslim …).

An **ordinal variable** consists of data that are categorized, and there is a clear order to the categories. Normally all scales are transformed into ordinal variables.

For example, educational experience is a typical variable to be grouped as an ordinal variable. For example, a coding in the following categories could make sense: no elementary school, elementary school graduate, high school graduate, college graduate, master's degree, Ph.D.

Table 4.11 Ordinal coding of the variable time it takes somebody to go to work

Less than 10 min	10-30 min	30-60 min	More than 60 min

Table 4.12 Ordinal coding of variable satisfaction with Chancellor Scholz foreign policy

1 not satisfied at all	2	3	4	5 very satisfied

An ordinal variable can also be a variable that categorizes a variable into set intervals. Such an interval question could be: How much time does it take you to get to work in the morning? Please tick the applicable category (see Table 4.11):

What is important for such an ordinal coding of a continuous variable is that the intervals are comparable and that there is some type of linear progression.

The most frequent types of ordinal variables are scales. Representative of such scales would be the answer to the question: Please indicate how satisfied you are with German Chancellor Scholz's foreign policy on a scale from 1 to 5 (see Table 4.12).

4.9 Coding of Different Variables in a Dataset

Except for string variables, question responses need to be transformed into numbers to be useful for data analytical purposes. Table 4.13 highlights some rules how this is normally done:

1. In the first column in Table 4.13, we have a string variable. This first variable identifies the participants of the study. In a real scientific study, the researcher would render the names anonymous and write "student 1, student 2...."
2. The second column is dichotomous variable (i.e., a variable with two possible response choices). Whenever researchers have a dichotomous variable, they create two categories labeling the first category 0 and the second category 1. (For any statistical analysis, it does not matter which one of the two categories you code 0 and 1, but, of course, you need to remember your coding to interpret the data correctly later).

Table 4.13 Representing string, dichotomous, continuous, and ordinal variables in a dataset

Student	Gender	Income	Age	Job satisfaction
John	1	12,954	43	2
Mary	0	33,456	67	1
James	1	98,534	54	0

3. Columns three and four are continuous variables representing the self-reported income and self-reported age the respondent has given in the questionnaire.
4. The fourth column captures a three-item ordinal variable asking individuals about their satisfaction with their job. The choices respondents are offered are low satisfaction (coded 0), medium satisfaction (coded 1), and high satisfaction (coded 2). Normally, ordinal variables are consecutively labelled from 0 or 1 for the lowest category to how ever many categories there are. To code ordinal variables, it is important that categories are mutually exclusive (i.e., one value cannot be in two categories). It is also important that progression between the various categories is linear.

4.9.1 Coding of Nominal Variables

For some analysis such as regression analysis (see Chaps. 8 and 9), non-dichotomous categorical variables (e.g., different religious affiliations) cannot be expressed in any ordered form, because there is no natural progression in their values. For all statistical tests that assume progression between the categories this causes a problem. To circumvent this problem for these tests, we use dummy variables to express such a variable. The rule for the creation of dummy variables is that we create one dummy variable less, than we have categories). For example, if we have four categories, we create three dummy variables, with the first category serving what is called the reference category. In the example in Table 4.14, the Muslim religion is the reference category against which we compare the other religions. (Since this procedure is rather complex, we will discuss the creation of dummy variables in Chap. 5 again).

4.10 Drafting a Questionnaire: General Information

In real research, the selection of an overarching question guiding the research process and the development of a questionnaire is theory driven. In other words, social science theory should inform researchers' choices of survey questions, and a good survey should respond to a precise theoretically driven research question (Mark, 1996: 15f). Yet, for the exercise in this book, where you are supposed to create your own questionnaire, you are not supposed to be an expert in a particular

Table 4.14 Representing a nominal variable in a dataset

	Dummy1	Dummy2	Dummy3
Muslim	0	0	0
Christian	1	0	0
Buddhist	0	1	0
Hindu	0	0	1

field. Rather, you can use your general interest, knowledge, and intuition to draft a questionnaire. Please also keep in mind that you can use your personal experience to come up with the topic and overarching research question of your sample survey. To highlight, if a researcher has been affected by poverty in his childhood, he might intuitively have an idea about the consequences of child poverty. In her case, it is also likely that she has read about this phenomenon and liked a hypothesis of one of the authors, she has read (e.g., that people who suffered from poverty in their childhood are less likely to go to college). Once you have identified a topic and determined the goals and objectives of your study, think about the questions you might want to include in your survey. Locating previously conducted surveys on similar topics might be an additional step enabling you to discover examples of different research designs about the same topic.

When you select the subject of your study it is important to take the subject's practicability into account as well. For one, a study on the political participation of university students can be relatively easily carried through by a university professor or a student. On the other hand, studies on human behavior in war situations for instance are very difficult to be carried out (Schnell et al., 2011: 3f.). Also keep in mind that before thinking about the survey questions, you must have identified dependent, independent, and control variables. You must have a concrete idea how these variables can be measured in your survey. For example, if a researcher wants to explain variation in student's grades, a student's cumulative grade average is the dependent variable. Possible independent or explanatory variables are the numbers of hours a student studies per week, the level of her class attendance gauged in percent of all classes, her interest in her field of study gauged from 0 not interested at all to 10 very interested, and her general life satisfaction again measured on a 0 to 10 scale. In addition, she might add questions about the gender, place of residency (city, suburban area or countryside), and the year of study (i.e., freshman, sophomore, junior, senior, or first, second, third, and fourth year).

4.10.1 Drafting a Questionnaire: Step-By-Step Approach

1. Think of an interesting social science topic of your choice (be creative), something that is simple enough to ask fellow students or your peers.
2. Specify the dependent variable as a continuous variable.[1]
3. Think of 6 to 8 independent (explanatory) variables which might affect the dependent variable. You should include three types of independent variables: continuous-, ordinal- and nominal/dichotomous variables. Do not ask questions that are too sensitive and do not include open-ended questions.

[1] For data analytical reasons, it is important to specify your dependent variable as a continuous variable, as the statistical techniques, you will learn later frequently require that the dependent variable is continuous.

4. On the question sheet, clearly label your dependent variable and your independent variables.
5. On a separate sheet, add a short explanation where you justify your study's topic. Also add several sentences per independent variable where you formulate a hypothesis and very quickly justify your hypothesis. In justifying your hypothesis, you do not need to consult the academic literature. Rather you can use your intuition and knowledge of the work you have already read.
6. If you encounter problems specifying and explaining a hypothesis, the survey question you have chosen might be suboptimal and you may want to choose another research question.
7. Add a very short introduction to your survey, where you introduce your topic. The example below can serve as a basis when you construct your survey.

4.11 Example of Questionnaire

Dear participants,
This survey is about the money college students spend partying, and possible factors that might influence students' partying expenses. Please answer the following questions which will be exclusively used for data analytical purposes in my political science research methods' class. If you are not sure about an answer, just give an estimate. All answers will be treated confidentially. We thank you for contributing to our study.

Dependent variable

- How much money do you spend partying per week?

Independent variables

- What is your gender?

  ☐ Male ☐ Female

- What is your year of study? (Circle one)

  1 2 3 4 5 6 or higher

- On average, how many hours do you spend studying per week?
- On average, how on how many days do you go partying per week? (Circle one)

  1 2 3 4 5 or more

- On a scale from 0 to 100 determine how much fun you can have without alcohol (0 meaning you can have a lot of fun; 100 you cannot have fun at all)?
- On a scale from 0 to 100 determine the quality of the officially sanctioned free-time activities at your institution (0 meaning they are horrible; 100 meaning they are very good)
- What percent of your tuition do you pay yourself?

4.11.1 Background Information About the Questionnaire

Study Purpose: This research analyzes university students' spending patterns, when they go out to party. This information might be important for university administrations, businesses, and parents. It might also serve as an indication of how serious they take their studies. Finally, it gives us some measurement on how much money students have at their disposal and how they spend it (or at least part of it).

Hypotheses

H(1): Male students spend more money partying than female students.

It is highly possible that male students like the bar scene more than female students do. For one, they go there to watch sports events such as soccer games. Additionally, they like to hang out there with friends while having some pints.

H(2): The more advanced students are in their university career, the less money they spend going out.

More advanced classes are supposedly getting more difficult than first- or second-year classes. This would imply that students must spend more time studying and would therefore have less time to go out. This, in turn, would entail that they are likely to spend less money for partying purposes.

H(3): The more time students spend studying, the less money they spend going out,

The rationale for this hypothesis is that students that study long hours just do not have the time to go out a lot and hence are unlikely to spend a lot of money.

H(4): The more students think they need alcohol to have fun, the more they will spend going out.

Alcohol is expensive, and the more students drink while going out the more money they will spend at bars and nightclubs.

H(5): The more students think that their university offers them good free time activities, the less money they will spend while going out.

If students spend a lot of their free time participating in university sponsored sports or social clubs, they will not have the time to go to bars or nightclubs frequently. As a result, they will not spend significant amounts of money while going out.

H(6): People that pay their tuition in full or partly will not spend as much money at bars than people that have their tuition paid.

The rationale for this hypothesis is straightforward: students that must pay a significant amount of their tuition might not have a lot of money left to spend at bars and nightclubs.

References

Bowling, A. (1997). *Research methods in health.* Open University Press.

Burns, N., & Grove, S. K. (1997). *The practice of nursing research conduct, critique, & utilization.* W.B. Saunders and Co.

Hoffman, A. (2014). *Indirekte Befragungstechniken zur Kontrolle sozialer Erwünschtheit in Umfragen.* Doctoral Thesis, Düsseldorf

Krosnick, J. A. (1999). *Maximizing questionnaire quality.* In: J. P. Robinson, P. R. Shaver, & L. S. Wrightsman (Eds.), *Measures of political attitudes: Volume 2 in measures of social psychological attitudes series.* Academic Press.

Kumar, R. (1999). Research methodology: A Step-by-step guide for beginners. In: E. D. De Leeuw, J. J. Hox, & D. A. Dillman (Ed.), *International handbook of survey methodology.* Lawrence Erlbaum Associates.

Mark, R. (1996). *Research made simple: A Handbook for social workers.* Sage Publications.

Mondak, J., & Davis, B. C. (2001). Asked and answered: Knowledge levels when we will not take "don't know" for an answer. *Political Behaviour, 23*(3), 199–224.

Neuman, W. L., & Robson, K. (2014). *Basics of social research.* Pearson Canada.

Sapsford, R. (2006). *Survey research.* Sage.

Schnell, R., Hill, P. B., & Esser, E. (2011). *Methoden der empirischen Sozialforschung* (9th ed.). Oldenbourg.

Schumann, S. (2012). *Repräsentative Umfrage: Praxisorientierte Einführung in empirische Methoden und statistische Analyseverfahren* (6th ed.). Oldenbourg.

Seidman, I. (2013). *Interviewing as qualitative research: A guide for researchers in education and the social sciences.* Teachers College Press.

Steenkamp, J.-B., De Jong, M. G., & Baumgartner, H. (2010). Socially desirable response tendencies in survey research. *Journal of Marketing Research, 47*(2), 199–214.

Sturgis, P., Allum, N., & Smith, P. (2008). An experiment on the measurement of political knowledge in surveys. *Public Opinion Quarterly, 72*(1), 90–102.

Van de Vijver, F. J. R., & He, J. (2014). *Report on social desirability, midpoint and extreme responding in TALIS 2013.* OECD education working papers, No. 107. OECD Publishing.

Weisberg, H. F., Krosnick, J. A., & Bowen, B. D. (1996). *An introduction to survey research, polling, and data analysis* (3rd ed.). Sage Publications.

Yin, R. K. (2015). *Qualitative research from start to finish.* Guilford Publications.

Further Reading

Nuts and Bolts of Survey Research

Nardi, P. M. (2018). *Doing survey research: A guide to quantitative methods*. Routledge (Chap. 1; Chap. 4).

Chapter 1 of this book provides a nice introduction why we do survey research, why it is important and what important insights it can bring to the social sciences. Chapter 4 gives a nice introduction into questionnaire developments and the different steps that go into the construction of a survey.

Constructing a Survey

Krosnick, J. A. (2018). Questionnaire design. In *The Palgrave handbook of survey research* (pp. 439–455). Palgrave Macmillan, Cham.

Short and comprehensive summary of the dominant literature into questionnaire design. Good as a first read of the topic.

Saris, W. E., & Gallhofer, I. N. (2014). Design, evaluation, and analysis of questionnaires for survey research. Wiley & Sons.

A very comprehensive guide into the design of surveys. Among others, the book thoroughly discusses how concepts become questions, how we can come up with response categories for the questions we use, and how to structure questions.

Applied Texts: Question Wording

Lundmark, S., Gilljam, M., & Dahlberg, S. (2015). Measuring generalized trust: An examination of question wording and the number of scale points. *Public Opinion Quarterly*, *80*(1), 26–43.

The authors show that the question wording of the general questions about trust in other individuals (i.e., generally speaking, would you say that most people can be trusted?) matters in getting accurate responses.

Conducting a Survey 5

5.1 Population and Sample

When the questionnaire is in its final form, the researcher needs to determine what the population of her study is and what type of sample she will take (see Fig. 5.1).

The **population** is the entire group of subjects the researcher wants information on. Normally the researcher or polling firm is not able to interview all units of the population because of the sheer size. To highlight, the United States has over 300 million inhabitants, Germany over 80 million and France over 65 million. It is logistically and financially impossible to interview the whole population. Therefore, the pollster needs to select a **sample** of the population instead. To do so, she first needs to define a **sampling frame**. A sampling frame consists of all units from which the sample will be drawn. Ideally, the sample frame should be identical to the population or at least closely resemble it. In reality, population and sampling frame frequently differ (Weisberg, 1996:39). For example, let us consider that the population are all inhabitants of Berlin. A reasonable sampling frame would include all households in the German capital, from which a random sample could be taken. Using this sample frame, a researcher could then send a questionnaire or survey to these randomly chosen households. However, this sampling frame does not include homeless people; because they do not have a fixed address, they cannot receive the questionnaires. Consequently, this sample drawn from the population will be slightly biased, as it will not include the thousands of people, who live on the streets in Berlin.

A **sample** is a subset of the population the researcher actually examines to gather her data. The collected data on the sample aims at gaining information on the entire population (Bickmann & Rog, 1998: 102). For example, if the German government wants to know whether individuals favor the introduction of a highway usage fee in Germany, it could ask 1000 people whether or not they agree with this proposal. For this survey, the population is the 82 million habitants of Germany, and the sample is the 1000 people, which the government asks. To make

© The Author(s), under exclusive license to Springer Nature Switzerland AG 2023

D. Stockemer and J.-N. Bordeleau, *Quantitative Methods for the Social Sciences*, Springer Texts in Political Science and International Relations,
https://doi.org/10.1007/978-3-031-34583-8_5

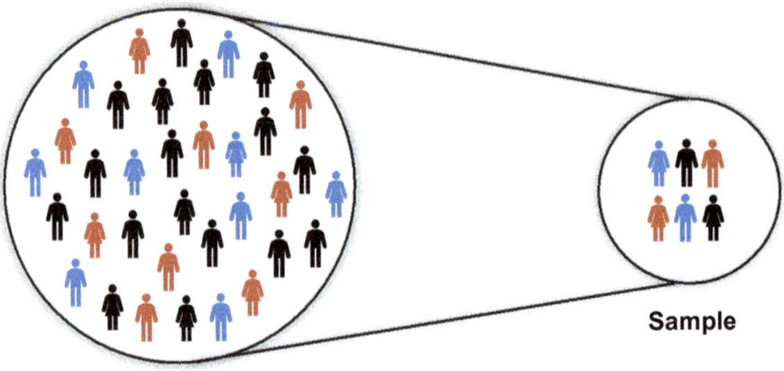

Fig. 5.1 Graphical display of a population and a sample

valid inference from a sample for a whole population the sample should be either representative or random.

5.2 Representative, Random, and Biased Samples

Representative Sample: a representative sample is a sample in which the people in the sample have the same characteristics as the people in the population. For example, if a researcher knows that in the population she wishes to study 55% of people are men, 18% are Afro-Americans, 7% are homeless and 23% earn more than 100,000 Euros, she should try to match these characteristics in the sample in order to represent the population.

Random Sample: in many social settings it is basically impossible for researchers to match the population characteristics in the sample. Rather than trying any matching technique, researchers can take a random sample. Randomization helps to offset the confounding effects of known and unknown factors by randomly choosing cases. For example, the lottery is a random draw of 6 numbers between 1 and 49. Similarly large-scale international surveys (e.g., the European Social Survey) use randomization techniques to select participants (Nachmias & Nachmias, 2008). Ideally, such randomization techniques give every individual in the population the same chance to be selected in the sample.

Biased Sample: a biased sample is a sample that is neither representative nor random. Rather than being a snapshot of the population, a biased sample is a sample, whose answers do not reflect the answers we would get had we the possibility to poll the whole population.

There are different forms of biases survey responses can suffer from:

Selection bias: We have selection bias if the sample is not representative of the population it should represent. In other words, a sample is biased if some type of individuals such as middle-class men are overrepresented and other types such as unemployed women are underrepresented. The more this is the case, the more biased the sample becomes and the potentially more biased the responses will be. Much of the early opinion polls that were conducted in the early twentieth century were biased. For example, the aforementioned Literary Digest poll, which was sent to around 10 million people prior to the Presidential Elections 1916 to 1936, was a poll that suffered from serious selection bias. Even though it correctly predicted the presidential winners of 1916 to 1932 (but it failed to predict the 1936 winner), it did not represent the American voting population accurately. To mail out its questionnaire the Literary Digest used three sources, its own readership, registered automobile users and registered telephone users. Yet, the readers of the Literary Digest were middle- or upper-class individuals, and so were automobile and telephone owners. For sure, in the 1910s, 20s and 30s, the use of these sources was probably a convenient way to reach millions of American. Nevertheless, the sampling frame failed to reach poor working-class Americans, as well as the unemployed. During the Great Depression in 1936 rather few Americans could afford the luxuries of reading a literary magazine or owning a car or a telephone. Hence, the unrepresentativeness of the Literary Digest's sample was probable aggravated in 1936. To a large degree, this can explain why the Literary Digest survey predicted Alfred Landon to win the presidential election in 1936, whereas Franklin D. Roosevelt won in a landslide amassing 61% of the popular vote. In fact, for the Literary Digest poll, the bias was aggravated by non-response bias (Squire, 1988).

Non-response bias: Non-response bias occurs, if certain individuals in your sample have a higher likelihood to respond than others, and if the responses of those, who do not respond would differ considerably from the responses of those, who respond. The source of this type of bias is self-selection bias. For most surveys (except for the census in some countries), respondents normally have their entirely free-will to decide whether or not to participate in the survey. Naturally, some people are more likely to participate than others are. Most frequently, this self-selection bias stems from the topic of the survey. To highlight, individuals, who are politically interested and knowledgeable might be more prone to answer a survey on conventional and unconventional political participation than individuals, who could not care less about politics. Yet, non-response bias could also stem from other sources. For example, it could stem from the time that somebody has at her disposal, as well as from access to technology. Persons with a busy professional and private life might simply forget to either fill out or return a survey. In contrast, individuals with more free time might be less likely to forget to fill out the survey; they might also be more thorough in filling it out. Finally, somebody's likelihood to fill out a survey might also be linked to technology (especially for online surveys). For example, not everybody has a smart phone or permanent access to the internet, some people differ in their email security settings, and some people might just not regularly check their emails. This implies that a survey might reach some people of the initial sample, but probably not all of them (especially if it is a survey that was sent out by email).

Response bias: Response bias happens when respondents answer a question misleadingly or untruthfully. The most common form of response bias is the so-called social desirability bias, respondents might try to answer the questions less according to their own convictions, but more in an attempt to adhere to social norms (see also Sect. 4.5). For example, social or behavioral surveys (such as the European Social Survey or National Election Studies) frequently suffer from response bias for the simple question whether individuals voted or not. Voting is an act that is socially desirable; a good citizen is expected to vote in an election. When asked the simple question whether or not they voted in the past national, regional or any other election, citizens know this social convention. Some non-voters might feel uneasy to admit they did not vote and indicate in the survey that they cast their ballot, even if they did not do so. In fact, over-reporting in election surveys is about 10–15 percentage points in Western democracies (see: Zeglovitz & Kritzinger, 2014). In contrast to the persistent over-reporting of electoral participation, the vote share for radical right-wing parties is frequently under-reported in surveys. Parties like the Front National in France or the Austrian Freedom Party attract voters and followers with their populist, anti-immigration and anti-elite platforms (Mudde & Kaltwasser, 2012). Voting for such a party might involve some negative stigmatisation as these parties are shunned by the mainstream political elites, intellectuals and the media. For these reasons, citizens might feel uneasy divulging their vote decision a survey. In fact, the real percentage of citizens voting for a radical right-wing party is sometimes twice as high as the self-reported vote choice in election surveys (see Stockemer, 2012).

Response bias also frequently occurs for personality traits and for the description of certain behaviors. For example, when asked, few people are willing to admit that they are lazy or that they chew their fingernails. The same applies to risky and illegal behaviors. Few individuals will openly admit engaging in drug consumption or will indicate in a survey that they have committed a crime.

Biased samples are ubiquitous in the survey research landscape. Nearly any freely accessible survey you find in a magazine or on the internet has a biased sample. Such a sample is biased because not all the people from the population see the sample, and of those, who see it not everybody will have the same likelihood to respond. In many freely accessible surveys, there is frequently also the likelihood to respond several times. A blatant example of a biased sample would be the National Gun Owner's Action Survey 2018 conducted by the influential US gun lobbyist, the National Rifle Association (NRA). The survey consists of 10 value-laden questions about gun rights. For example, in the survey, respondents are asked: should Congress and the states eliminate so-called "gun free zones" that leave innocent citizens defenceless against terrorists and violent criminals? The anti-gun media claims that most gun owners support mandatory, national gun registration? Do you agree that law-abiding citizens should be forced to submit to mandatory gun registration or else forfeit their guns and their freedom? In addition to these value laden questions, the sample is mainly restricted to NRA members, gun owners who cherish the second amendment of the American Constitution. Consequently, they will show high opposition toward gun control. Yet, their opinion will certainly not be representative of the American population.

Yet, surveys are not only (ab)used by think tanks and non-governmental organisations to push their demands, in recent times, political parties and candidates also us (online) surveys for political expediency or as a political stunt. For example, after taking office in January 2017, president's Trump campaign sent out an online survey to his supporters entitled "Mainstream Media Accountability Survey". In the introduction the survey directly addressed the American people—"you are our last line of defense against the media's hit jobs. You are our greatest asset in helping our movement deliver the truth to the American people." The questionnaire then asked questions like: "has the mainstream media reported unfairly on our movement?" "Do you believe that the mainstream media does not do their due diligence of fact-checking before publishing stories on the Trump administration?" or "Do you believe that political correctness has created biased news coverage on both illegal immigration and radical Islamic terrorism?" From a survey perspective, Trump's survey is the anti-example of doing survey research. The survey pretends to address the American people, whereas in fact it is only sent to Trump's core supporters. It further uses biased and value-laden questions. The results of such a biased survey is a political stunt that the Trump campaign still uses for political purposes. In fact, with the proliferation of online surveys, with the continued discussion about fake news, and with the ease with which a survey can be created and distributed, there is the latent danger that organisations try to send out surveys less to get a "valid" opinion from a population or clearly defined subgroup of that population, but rather as a stunt to further their cause.

5.3 Sampling Error

For sure, the examples described under biased surveys have high sampling errors, but even with the most sophisticated randomization techniques, we can never have a 100% accurate representation of a population from a sample. Rather, there is always some statistical imprecision in the data. Having a completely random sample would imply that all individuals who are randomly chosen to participate in the survey actually do participate, something that will never happen. The sampling error depicts the degree to which the sample differs from a population. For a random sample, there is a formula of how to calculate the sampling error (see Sect. 6.6). Basically, the sampling error depends on the number of observations (i.e., the more observations we have in the sample, the more precision there is in the data) and the variability of the data (i.e., how much peoples' opinions differ). For example, if we ask Germans to rate chancellor Scholz's' popularity from 0 to

100, we will have a higher sampling error if we ask only 100 instead of 1000 individuals.[1] Similarly, we will have a higher sampling error, if individuals' opinions differ widely rather than being clustered around a specific value. To highlight, if Scholz's popularity values differ considerably—that is some individuals rate him at 0, others at 100—there is more sampling error than when nearly everybody rates him around 50, because there is just more variation in the data.

5.4 Non-random Sampling Techniques

Large-scale national surveys, measuring the popularity of politicians, citizens' support for a law, or citizens voting intentions generally use random sampling techniques. Yet, not all research questions require a random sample. Sometimes random sampling might not be possible or too expensive. The most common non-probabilistic sampling techniques are convenience sampling, purposive sampling, volunteer sampling, and snowball sampling,

Convenience sampling: Convenience sampling is a type of non-probabilistic sampling technique where people are selected because they are readily available. The primary selection criterion relates to the ease of obtaining a sample. One of the most common examples of convenience sampling is using student volunteers as subjects for research (Battaglia, 2008). In fact, college students are probably the most frequently used group in psychological research. For instance, many researchers (e.g., Praino et al., 2014) examining the influence of physical attractiveness on the electoral success of candidates for political office use college students from their local college or university to rank the physical attractiveness of their study subjects (i.e., political candidates running for office). These students are readily available and cheap to recruit.

Purposive sampling: In purposive sampling, subjects are selected because of some characteristics, which the researcher predetermines before the study. Purposive sampling can be very useful in situations where the researcher needs information for a specific target group (e.g., blond women aged 30 to 40). She can purposefully restrict her sample to the required social group. A common form of purposive sampling is expert sampling. An expert sample is a sample of experts with known and demonstrable expertise in a given area of interest. For example, a researcher uses an expert sampling technique, if she sends out a survey to corruption specialists to ask them about their opinions about the level of corruption in a country (Patton, 1990). In fact, most major international corruption indicators such as Transparency International, the World Bank Anti-Corruption Indictor or the electoral corruption indicators collected by the Electoral Integrity Project are all constructed from expert surveys.

[1] Increasing the number of participants in the sample increases the precision of the data up to a certain number such as 1000 or 2000 participants. Beyond that number, the gains in increasing precision are limited.

Volunteer sampling: Volunteer sampling is a sampling technique frequently used in psychology or marketing research. In this type of sampling volunteers are actively searched-for or invited to participate. Most of the internet surveys that flood the web also use volunteer sampling. Participants in volunteer samples often have an interest in the topic, or they participate in the survey because they are attracted by the money or the non-financial compensation they receive for their participation (Black, 1999). Sometimes pollsters also offer a high monetary prize for one or several lucky participants to increase participation. In our daily lives, volunteer surveys are ubiquitous, ranging from airline passenger feedback surveys to surveys about costumer habits, to personal hygiene questionnaires.

Snowball sampling: Snowball sampling is typically employed with populations, which are difficult to access. The snowball sampling technique is relatively straightforward. In the first step, the researcher has to identify one or several individuals of the group she wants to study. She then asks the first respondents if they know others of the same group. By continuing this process, the researcher slowly expands her sample of respondents (Spreen, 1992). For example, if a researcher wants to survey homeless people in the district of Kreuzberg in Berlin, she is quite unlikely to find a list with all the people who live in the street. However, if the researcher identifies several homeless people, they are likely to know other individuals that live in the streets, who again might know others.

Quota sampling: As implied in the word, quota sampling is a technique (which is frequently employed in online surveys), where sampling is done according to certain pre-established criteria. For example, many polls have an implicit quota. For example, customer satisfaction polls, membership polls and readership polls, all have an implicit quota. They are restricted to those that have purchased a product or service for customer satisfaction surveys, the members of an organisation or party for membership surveys, and the readers of a journal, magazine or online site for readership polls. Yet, quota sampling can also be deliberately used to increase the representativeness of the sample. For example, let us assume that a researcher wants to know how Americans think about same sex marriage. Let us further assume that the researcher sets the sample size at 1000 people. Using an online questionnaire, she cannot get a random or fully representative sample, because comparatively few people sign up to participate in online panels. Yet, what she can do is to make her sample representative of some characteristics such as gender and region. By setting up quotas, she can do this relatively easily. For example, as regions, she could identify the East, the Mid-West, the West and the South of the United States. For gender, she could split the sample so that she has 50% men and women. This gives her a quota of 125 men and 125 women for each region. Once, she reaches this quota, she closes the survey for this particular cohort (for the technical details on how this works, see also Fluid Survey University 2017). Using such a technique allows researchers to build samples that more or less reflect the population at least when it comes to certain characteristics. While it is cheaper than random sampling, quota sampling with the help of a survey company can still prove rather expensive. For example, using an online quota sampling for a short questionnaire on Germans' knowledge and assessment of the fall of the Berlin Wall in 1989, which the senior co-author of this book conducted in 2014, he paid $8 per stratified survey. The stratification criteria he used were first gender balance and second the requirement that half the participants must reside in the East of Germany and the other half in the West.

5.5 Different Types of Surveys

Questions about sampling and the means through which a survey is distributed often go hand in hand. Several sampling techniques lend themselves particularly well to one or another distribution medium. In survey research, we distinguish four different ways to conduct a survey: face-to-face surveys, telephone surveys, mail-in surveys, and online surveys.

Face-to-face surveys: historically the most commonly employed survey method is the face-to-face survey. In essence, in a face-to-face interview or survey, the interviewer travels to the respondent's location, or the two meet somewhere else. The key feature is the personal interaction between the interviewer who asks questions from a questionnaire and the respondent who answers the interviewer's questions. The direct personal contact is the key difference from a telephone interview, and it comes with both opportunities and risks. One of the greatest advantages is that the interviewer can also examine the interviewee's non-verbal behaviour and draw some conclusions from this. She can also immediately respondent when problems arise during the task performance; for example, if the respondent does not understand the content of a question the interviewer can explain the question in more detail.

Therefore, this type of survey is especially suitable when it comes to long surveys on more complex topics; topics where the interviewer must sit down with the respondent to explain certain items. Nevertheless, this type of survey also has its drawbacks: a great risk with face-to-face surveys stems from the impact of the interviewer's physical presence on the respondents' answers. For example, slight differences about the interviewers' ways of presenting an item can influence the responses. In addition, there is the problem of social desirability bias. In particular, in the presence of an interviewer respondents could feel more pressured to meet social norms when answering questions. For example, in the presence of a pollster individuals might be less likely to admit that they have voted for a radical right-wing party or that they support the death penalty. In a more anonymous survey, such as an online survey, the respondents might be more willing to admit socially frowned upon behaviors or opinions. Face-to-face surveys are still employed for large-scale national surveys such as the census in some countries. Some sampling techniques, such as snowball sampling, also work best with face-to-face surveys.

Telephone survey: in many regards, the telephone interview resembles the face-to-face interview. Rather than personal, the interaction between interviewer and interviewee is via the phone. Trained interviewers can ask the same questions to different respondents in a uniform manner thus fostering precision and accuracy in soliciting responses. The main difference between a telephone interview and a personal interview is logistics. Because the interviewer does not have to travel to the interviewee's residence or meet her in a public location, a larger benefit from this technique is cost. Large-scale telephone surveys significantly simplify the supervision of the interviewers as most or all of them conduct the interviews from the same site. More so than face-to-face surveys, telephone surveys can also take advantage of recent technological advancements. For example, so-called computer assisted telephone surveys (CATS) allow the interviewer to record the data directly into a computer. This has several advantages: (1) there is no need for any time-consuming transfer process from a transcription medium to a computer (which is a potential source of errors too). (2) The computer can check immediately whether given responses are invalid and change or skip questions depending on former answers. In particular, survey firms use telephone interviews extensively, thus benefiting from the accuracy and time efficiency of modern telephone surveys coupled with modern computer technology (Carr & Worth, 2001; Weisberg et al., 1996112–113).

Mail-in survey: mail-in surveys are surveys that are sent to peoples' mailboxes. The key difference between these self-administered questionnaires and the aforementioned methods is the complete absence of an interviewer. The respondent must cope with the questionnaire herself; she only sees the questions and does not hear them; assistance cannot be provided and there is nobody who can clarify unclear questions or words. For this reason, researchers or survey firms must devote great care when conceiving a mail in survey. In particular, question wording, the sequence of questions, and the layout of the questionnaire must be easy to understand for respondents (Leeuw & Hox, 2008: 239–241). Furthermore, reluctant "respondents" cannot be persuaded by an interviewer to participate in the survey, which results in relatively low response rates. Yet, individuals can take the surveys at their leisure and can think about an answer as much time as they like.

A potential problem of mail-in-surveys is the low response rate. Sometimes, only 5, 10 or 20% of the sample sends the questionnaire back, a feature which could render the results biased. To tackle the issue of low response rates the researcher should consider little incentives for participation (e.g., the chance to win a prize or some compensation for participation) and she should send follow-up mailings to increase the participants' willingness to participate. The upside of the absence of an interviewer is that the researcher does not need to worry about interviewer effects biasing the results. Another advantage of mail in surveys is the possibility to target participants. Provided that the researcher has demographic information about each household in the population or sample she wants to study, mail-in surveys allow researchers to target the type of individuals she is most interested in. For example, if a researcher wants to study the effect of disability on political participation, she could target only those individuals, who fall under the desired category, people with disabilities, provided she has the addresses of those individuals.

Online survey: online surveys are essentially a special form of mail in surveys. Instead of sending a questionnaire by mail, researchers send online surveys by email or use a website such as Survey Monkey to host their questionnaire. Online surveys have become more and more prominent over the past 20 years in research. The main advantage is costs. Thanks to survey sites such as Survey Monkey, or Fluid Survey, everybody can create a survey with little cost. This also implies that the use of online surveys is not restricted to research. To name a few, fashion or sports magazines, political parties, daily newspapers, as well as radio and television broadcasting stations all use online surveys. Most of the time, these surveys are open to the interested reader and there are no restrictions for participation. Consequently, the answers to such surveys frequently do not cater to the principles of representativeness. Rather, these surveys are based on a quota or convenience sample. Sometimes this poses little problems, as many online surveys target a specific population. For example, it frequently happens to researchers who publish in a scientific journal with publishing houses such as Springer or Tyler and Francis that they receive a questionnaire asking them to rate their level of satisfaction with the publishing experience with the specific publishing house. Despite the fact that the responses to these questionnaires are almost certainly unrepresentative of the whole population of scientists the feedback they receive might give these publishing houses an idea which authors' services work and which do not work, and what they can do to make the publishing experience more rewarding for authors. Yet, online surveys can also more problematic. They become particularly problematic if an online sample is drawn from a biased sample to draw inferences beyond the target group (see Sect. 5.2).

5.6 Which Type of Survey Should Researchers Use?

While none of the aforementioned survey types is a priori superior to the others, the type of survey researchers should use depends on several considerations. First, and most importantly, it depends on the purpose of the research and the researcher's priorities. For instance, many surveys, including surveys about voting intentions or the popularity of politicians, nearly by definition require some national random telephone, face-to-face or online samples. Second, surveys of a small subset of the population, who are difficult to reach by phone or mail, such as the homeless, similarly require face-to-face interactions. For such a survey, the sample will most likely be a non-random snowball or convenience sample. Third, for other surveys more relevant for marketing firms and companies, an online convenience sample might suffice to draw some "valid" inferences about customer satisfaction with a service or a product.

There are some more general guidelines. For one, online surveys can reach a large number of individuals at basically no cost, in particular if there are no quotas involved, and if the polling firm has the relevant email addresses or access to a hosting website. One the other hand, face-to-face and telephone interviews can target the respondents more thoroughly. If the response rate is a particularly important consideration, then personal face-to-face surveys or telephone surveys might also be a good choice. The drawback to these types of surveys is the cost; these personal surveys are the most expensive type of survey (with telephone surveys having somewhat lower costs than face-to-face surveys). For research questions where the representativeness or randomness of the sample is less of an issue, or for surveys that target a specific constituency, mail in or online surveys could be a rather cost-effective means. In particular, the latter are more and more frequently used, because through quota sampling techniques, these surveys can also generate rather representative samples, at least according to some characteristics. However, as the example of the Trump survey illustrates, online surveys can also be used for political expediency. That is why, the reader, before believing any of these survey results, in particular, in a non-scientific context, should inform herself about the sampling techniques, and she should look at question wording and the layout of the survey.

5.7 Pre-tests

5.7.1 What is a Pre-test?

Questionnaire design is complex; hardly any expert can design a perfect questionnaire by just sitting at her desk (Campanelli, 2008: 176). Therefore, before beginning the real polling a researcher should test her questions in an authentic setting to see if the survey as a whole and individual questions make sense and are easily understood by the respondents. To do so, she could conduct the survey with a small subset of the original sample or population aiming to minimize problems before the actual data collection begins (Krosnick, 1999: 50 f.; Krosnick & Presser, 2010: 266 f.) Such preliminary research or pre-tests make particularly sense in five cases.

First, for some questions researchers need to decide upon several possible measurements. For example, for questions using a Likert scale, a pre-test could show if most individuals choose the middle category or the do not know option. If this is the case, researchers might want to consider eliminating these options. In addition, for a question about somebody's income a pre-test could indicate whether respondents are more comfortable answering a question with set income brackets or if they are willing to reveal their real income.

Second, questions need to be culturally and situationally appropriate, easy to be understand, and they must carry the inherent concept's meaning. To highlight, if a researcher conducts a survey with members of a populist radical right-wing party, these members might feel alienated or insulted if she uses the word radical right or populist right. Rather, they define themselves as an alternative that incorporates common sense. Normally, a researcher engaging in this type of research should already know this, but in case he does not, a pre-test can alert her to such specificities allowing her to use situationally appropriate wording.

Third, a pre-test can help a researcher discover if responses to a specific item vary or not. In case there is no variance in the responses to a specific question at all, or very little variance, the researcher could think about omitting the issue to assure that the final questionnaire is solely comprised of discriminative items (items with variance) (Kumar, 1999: 132). For example, if a researcher asks the question whether the United States should leave NATO, and everybody responds with no, the researcher might consider dropping this question, as there is no variation in answers.

Fourth, a pre-test is particularly helpful if the survey contains open-ended questions, and if a coding scheme for these open-ended questions is developed alongside the survey. Regardless of the level of sophistication of the survey, there is always the possibility that unexpected and therefore unclassifiable responses will be encountered during the survey arise. Conducting a pre-test can reduce this risk.

Fifth, and more practically, a pre-test is especially recommendable if a group of interviewers (with little experience) run the interviews. In this case, the pre-test can be part of the interviewers' training. After the pre-test, the interviewers not only share their experiences and discuss which questions were too vague or ambiguous, but also share their experience asking the questions (Behnke et al., 2006: 258 f.).

After the pre-test, several questions might be changed depending on the respondents' reactions (e.g., low response rates) to the initial questionnaire. If significant changes are made in the aftermath of the pre-test, the revised questions should also be re-tested. This subsequent pre-test allows the researcher to check if the new questions, or the new question wordings, are clearer or if the changes have caused new problems (for more information on pre-testing see Guyette, 1983, p. 54–55).

5.7.2 How to Conduct a Pre-test?

The creation of a questionnaire/survey is a reiterative process. In a first step, the researcher should check the questions several times to see if there are any uncertain or vague questions, if the question flow is good and if the layout is clear and appealing. To this end, it also makes sense for the researcher to read the questions aloud so that she can reveal differences between written and spoken language. In a next step, she might run the survey with a friend or colleague. This preliminary testing might already allow the researcher to identify (some) ambiguous or sensitive questions or other problematic aspects in the questionnaire. Then, after this preliminary check, the researcher should conduct a trial or pre-test to verify if the topic of the questionnaire and every single question are well understood by the survey respondents. To conduct such a pre-test, researchers normally choose a handful of individuals who are like those that will actually take the survey. When conducting her trial, the researcher must also decide whether the interviewer (if he does not run the pre-test himself) informs the respondent about the purpose of the survey beforehand. An informed respondent could be more aware of the interviewing process and any kinds of issues that arise during the pre-test. However, the flipside is that the respondent may take the survey less seriously. The so-called respondent debriefing session offers a middle way addressing the described obstacles. In a first step, the pre-test is conducted without informing the respondent beforehand. Then, shortly after the end of the pre-test, the interviewer asks the respondent about obstacles and challenges the respondent encountered during the pre-test (Campanelli, 2008: 180).

References

Battaglia, M. (2008). Convenience sampling. In P. J. Lavrakas (Ed.), *Encyclopedia of survey research methods*. Sage Publications.

Behnke, J., Baur, N., & Behnke, N. (2006). *Empirische Methoden der Politikwissenschaft*. Schöningh.

Bickman, L., & Rog, D. J. (1998). *Handbook of applied social research methods*. Sage Publications.

Black, T. R. (1999). *Doing quantitative research in the social sciences: An integrated approach to research design, measurement, and statistics*. Sage Publications Inc.

Campanelli, P. (2008). Testing survey questions. In: E. D. De Leeuw, J. J. Hox, & D. A. Dillman (Eds.), *International handbook of survey methodology*. Lawrence Erlbaum Associates.

Carr, E. C., & Worth, A. (2001). The use of the telephone interview for research. *NT Research, 6*(1), 511–524.

De Leeuw, E. D., Hox, J. J., & Dillman, D. A. (2008). Mixed-mode surveys: When and why. *International handbook of survey methodology* (pp. 299–316). Routledge.

Fluid Surveys University. (2017). *Quota sampling effectively—How to get a representative sample for your online surveys*. Accessed on December 1st, 2017, from http://fluidsurveys.com/university/using-quotas-effectively-get-representative-sample-online-surveys/

Guyette, S. (1983). *Community based research: A handbook for native Americans*. American Indian Studies Center.

Krosnick, J. A. (1999). Maximizing questionnaire quality. In: J. P. Robinson, P. R. Shaver, & L. S. Wrightsman (Eds.), *Measures of political attitudes: Volume 2 in measures of social psychological attitudes series*. Academic Press.

Krosnick, J. A., & Presser, S. (2010). Question and questionnaire design. In: P. V. Marsden, & J. D. Wright (Eds.), *Handbook of survey research*. Emerald Group Publishing.

Kumar, R. (1999). Research methodology: A step-by-step guide for beginners. In: E. D. De Leeuw, J. J. Hox, & D. A. Dillman (Eds.), *International handbook of survey methodology*. Lawrence Erlbaum Associates.

Mudde, C., & Kaltwasser, C. R. (Eds.). (2012). *Populism in Europe and the Americas: Threat or corrective for democracy?* Cambridge University Press.

Nachmias, C. F., & Nachmias, D. (2008). *Research methods in the social sciences* (7th ed.). Worth.

Patton, M. Q. (1990). *Qualitative evaluation and research methods* (2nd ed.). Sage Publications.

Praino, R., Stockemer, D., & Moscardelli, V. G. (2013). The lingering effect of scandals in congressional elections: Incumbents, challengers, and voters. *Social Science Quarterly, 94*(4), 1045–1061.

Spreen, M. (1992). Rare populations, hidden populations and link-tracing designs: What and why? *Bulletin Methodologie Sociologique, 36*(1), 34–58.

Squire, P. (1988). Why the 1936 literary digest poll failed. *Public Opinion Quarterly, 52*(1), 125–133.

Stockemer, D. (2012). The Swiss radical right: Who are the (new) voters of Swiss peoples' party? *Representation, 48*(2), 197–208.

Weisberg, H. F., Krosnick, J. A., & Bowen, B. D. (1996). *An introduction to survey research, polling, and data analysis* (3rd ed.). Sage Publications.

Zeglovits, E., & Kritzinger, S. (2014). New attempts to reduce overreporting of voter turnout and their effects. *International Journal of Public Opinion Research, 26*(2), 224–234.

Further Reading

Constructing and Conducting a Survey

Kelley, K., Clark, B., Brown, V., & Sitzia, J. (2003). Good practice in the conduct and reporting of survey research. *International Journal for Quality in Health Care, 15*(3), 261–266.

The short article provides a hands-on step-by-step approach into data collection, data analysis, and reporting. For each step of the survey process, it identifies best practices and pitfalls to be avoided so that the survey becomes valid and credible.

Krosnick, J. A., Presser, S., Fealing, K. H., Ruggles, S., & Vannette, D. L. (2015). *The future of survey research: Challenges and opportunities*. The National Science Foundation Advisory Committee for the Social, Behavioral and Economic Sciences Subcommittee on Advancing SBE Survey Research. Available online at: http://www.nsf.gov/sbe/AC_Materials/The_Future_of_Survey_Research.pdf.

Comprehensive reports on the best practices, challenges, innovations, and new data-collection strategies in survey research.

Rea, L. M., & Parker, R. A. (2014). *Designing and conducting survey research: A comprehensive guide*. Wiley & Sons. A very comprehensive book into survey research consisting of three parts: (1) Developing and administering a questionnaire, (2) ensuring scientific accuracy, and (3) presenting and analyzing survey results.

Internet Survey

Alessi, E. J., & Martin, J. I. (2010). Conducting an internet-based survey: Benefits, pitfalls, and lessons learned. *Social Work Research, 34*(2), 122–128.

This text provides a very hand on introduction into the conduct of an internet survey with a special focus on recruitment strategies and response rate.

De Bruijne, M., & Wijnant, A. (2014). Improving response rates and questionnaire design for mobile web surveys. *Public Opinion Quarterly, 78*(4), 951–962.

This research note provides some practical lessons how the response rate and data quality can be improved for internet surveys especially constructed for smart phones.

Introducing R and Univariate Statistics 6

6.1 R Programming Language

R is a programming language that is gaining in popularity in the social sciences. Many political science departments in some of the largest universities have switched to teaching R. The conversion from traditional statistical packages like Stata and SPSS to R is taking place for many reasons. The number one reason for the surge of R as a data science tool is its accessibility. R and RStudio are open-source and therefore completely free to use. Unlike Stata and SPSS, which can cost several thousands of dollars, R can be downloaded for free on all computer systems. Moreover, R is an open-source programming language which operates through packages. Users from around the world can contribute to R by building new packages or improving existing ones. Contrary to traditional statistical software which often require updates (which can incur additional fees), R is in constant evolution and improvement by a variety of highly skilled users, data scientists, and statisticians. Before diving into examples of how we can analyze survey data using R, we must first cover how to install R and explore the RStudio interface.

6.1.1 Downloading R and RStudio

The first step in using R for survey data analysis is downloading the language and its interface RStudio. Let us explain the difference between these two essential elements. R is the programming language. However, without an interface to communicate, that language is not very useful. That is why RStudio exists. This software serves as an interface onto which researchers and data scientists use the R programming language to analyze data (think of RStudio as the mouth; without it you would not be able to speak R).

© The Author(s), under exclusive license to Springer Nature Switzerland AG 2023

D. Stockemer and J.-N. Bordeleau, *Quantitative Methods for the Social Sciences*, Springer Texts in Political Science and International Relations,
https://doi.org/10.1007/978-3-031-34583-8_6

In order to download R, you must follow this link (https://cran.r-project.org) and select the R version that is appropriate for your operating system (MacOS, Windows, or Linux). For the purpose of this book, we will use the MacOS (Apple) version of R for the examples.[1] However, Windows users must download Rtools (https://cran.r-project.org/bin/windows/Rtools/) which allows users to use certain statistical functions when using RStudio on Windows. Once R is downloaded (and Rtools for Windows users), you will need to download the RStudio interface. To do so, you must visit the following link (https://posit.co/downloads/) and download the RStudio Desktop Open Source Edition (AGPL). Now that you have downloaded both R and RStudio, it is time to familiarize ourselves with the basic R syntax and the RStudio interface.

6.1.2 RStudio Interface

The RStudio interface is divided into quadrants (see Fig. 6.1). The upper left quadrant is where users input their commands on an R Script. Scripts are equivalent to Syntax files in Stata: they present the code necessary to produce the analyses required by the user. These scripts can be saved and shared to be used by other users. This is especially useful to conduct verification of analyses and replication of studies. To put it simply, the command quadrant is where we need to write our code.

The lower left quadrant is where the output of our commands will be presented (with the exception of graphics and plots). For example, when we input a regression equation in the command and run the code, the results of the regression will be printed in the lower left quadrant. This is also where error messages will be printed. This quadrant can also be used as a direct input source for commands. If we want to run a quick command that we do not need to keep documented in our script, we can simply enter that command in the console and hit enter.

The upper right quadrant presents the R environment. This is where all the different datasets, variables, models, objects, and plots will be stored. In a certain way, this quadrant offers a list of the various objects that are present in a user's given R session. If we input a dataset, it will appear in this quadrant since it is a part of the environment we are using. Creating new variables will make them appear in this list, as well.

Lastly, the lower right quadrant is where plots and graphs will be printed. You can use this quadrant to export a given graph. You can also adjust the size of the quadrant to make a plot bigger or smaller. This is also where we can find various

[1] There is no difference in the R programming language between the different operating systems. However, the syntax for file paths will differ slightly since Windows and MacOS (Apple) use distinct file paths.

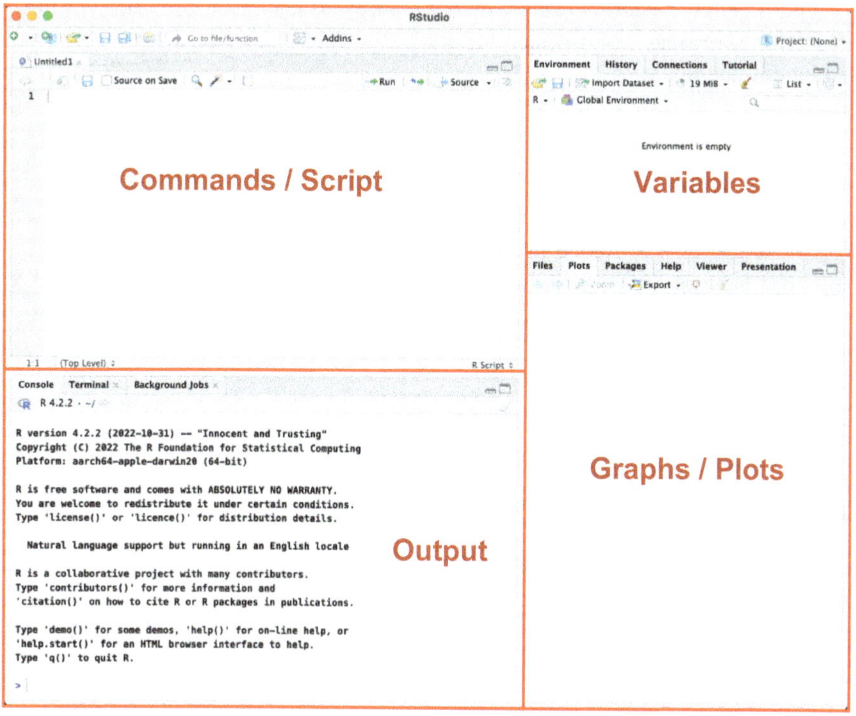

Fig. 6.1 RStudio interface

packages from the system library and get some help through instruction manuals. Entering a help command in the console will print the designated instruction manual in this quadrant.

6.1.3 R Packages

As we have previously mentioned, R works through packages. In other words, you need to download specific packages to make available certain commands and be able to execute certain models or analyses. By default, R has a set of basic commands. However, one of the most useful aspects of R are the packages built by the community which are constantly developed and improved. For example, the tidyverse package provides many useful commands to clean and tidy your data (e.g., data transformation and recoding) (Wickham et al., 2019). Another package which is useful is modelsummary. It allows users to create easy-to-read tables and present the results of all sorts of models (Arel-Bundock, 2022).

The install.packages() command is used to download packages and the library() command loads packages into your current R session. These two commands are essential to using R for data science. For every new R session, it is necessary to load the required packages (not re-install, simply re-load with the library() function). The following code installs and loads four packages that we will be using throughout this book:

```
install.packages("tidyverse")
install.packages("dplyr")
install.packages("modelsummary")
install.packages("foreign")

library(tidyverse)
library(dplyr)
library(modelsummary)
library(foreign)
```

The CRAN website keeps an updated list of all packages available. You can consult this list here:

https://cran.r-project.org/web/packages/available_packages_by_name.html.

6.1.4 The Basics of R

R works through variables, vectors, and data frames. Variables are objects for which values have been assigned using the <- operator. We can assign numerical, character, or even logical values to variables. For example, the line of code x<- 2 assigns the value of 2 to the variable x. A vector is created when combining a series of values. A vector is therefore a variable with multiple values or observations. A vector is built using the c() command. The following code, for instance, creates a vector:

```
## Let's create a vector 'x' with the values 1, 2, and 3.
x<- c(1, 2, 3)

## Now let's create a vector 'y' with the values "yes", "no", "maybe."
y<- c("yes", "no", "maybe")
```

When we bring vectors together, we create a data frame. Data frames are essentially a collection of data in the form of a dataset. It resembles an Excel spreadsheet with rows and columns. We can use the function data.frame() to create a data frame from specified vectors. Data frames and variables will appear in the upper-right quadrant in RStudio. We can click on the name of a data frame for it to open in a new window or simply type the name of the data frame in the console to view it. We can use the dollar sign operator $ to specify a vector we wish to use within a specific data frame (a column). Consider the following example:

```
## Let's create a data frame with the two vectors we previously generated.
dat < data.frame(x, y)

## We now have a data frame named 'dat' which includes vectors x and y.
dat
  x    y
1 1  yes
2 2   no
3 3 maybe

## We can single out the second column (vector y) using the $ operator
dat$y
[1] "yes" "no" "maybe"
```

If a line of code does not work as intended, R will print an error code in the output window. Usually, the error code will clearly indicate what went wrong so you can modify the code and fix the issue. Let us generate an error code by trying to assign a value to a variable using an inexistant variable.

```
## Let's try to assign the value of variable W to the variable Z
Z <- W
Error: object 'W' not found

## As we can see, we cannot assign the value of W to Z since W does not
exist.
```

Lastly, we can use a question mark in front of specific command names to get help on how to use the command. For example, if we enter the command ?install.packages the help section for this specific command will appear in the lower right quadrant of RStudio. In help sections, you can get information on the different arguments possible as well as the general function of the command.

6.2 Importing Data into R

Importing data in R is easy. Before doing so, however, we need to set our working directory. We can obtain the current working directory using the function getwd(). The working directory is the way by which we tell R where to get files, but also where to export files. In other words, it is where R will operate within your computer. We can set our working directory by specifying the file path to the location we desire:

```
## Let's set our working directory to our desktop folder.
setwd("/Users/Username/Desktop")

## To check if the command worked, let's check what our working directory
is.
getwd()
[1] "/Users/Username/Desktop"
```

Now that our working directory is set to the location where our data files are, we can proceed. It is possible to import several different types of data files including Excel spreadsheets, Stata data files, and comma-separated values files (.csv). The command we use to import data frames into R depends on the file extension of the dataset we want to import. The most common types of files we will use are .csv and .dta (comma-separated values and Stata data files). We can import .csv files using base R, but we need the foreign package to access the command to import .dta files. Remember, we have already installed and loaded this package in Sect. 6.1.3.

The commands read.csv() and read.dta() allow users to import datasets into R. When inputting the name of the file, it is important to place the name in between quotation marks (i.e., "filename.csv"). We can also import files directly from web addresses by replacing the file names with the link for the web address where the file is located. We will import a data file from one of our own online repositories to use for the example in this chapter.[2]

```
## Let's import the "PartyData.csv" data file from the web.
dat <- read.csv("https://raw.githubusercontent.com/nickbordeleau/
                                    QMSS/main/PartyData.csv")
View(dat)
```

As we can see, the data is now available as a data frame in our R session. We should be able to see eight variables (columns) and 40 observations (rows). This dataset contains data regarding the partying habits of 40 undergraduate students. The eight variables are labeled as follows: MSP for money spent partying, ST for study time, TP for times partying per week, FWA for fun without alcohol, QECA for quality of extracurricular activities, and ATSP for amount of tuition paid by the student. We will now explore univariate statistics and various data visualization tools using this data frame as an example.

[2] If you are having difficulty importing the data from our online repository, you can download the dataset to your computer and import it directly. To do so, visit the repository here: https://github.com/nickbordeleau/QMSS.

6.3 Frequency Table

A frequency table is a rather simple univariate statistic that is particularly useful for categorical variables such as ordinal variables that do not have too many categories. For each category or value, such a table indicates the number of times or percentage of times each value occurs. A frequency table normally has four columns. The first column lists the available categories. The second column displays the raw frequency or the number of times a single value occurs. The third column shows the percentage of observations that fall into each category—the basic formula to calculate this percentage is the number of observations in the category divided by the total number of observations. The final column, labeled cumulative percentage, displays the percentage of individuals up to a certain category or point.

Table 6.1 displays a frequency table for the variable TP. The first column displays the available options (i.e., the six categories ranging from zero to five times and more). The second column shows the raw frequencies (i.e., how many students indicates partying zero times, once, twice, three times, four times, or five times per week). The third column displays the raw frequency in percentages. The final column displays the cumulative percentage. For example, Table 6.1 highlights that 60% of the polled, on average, party two times or fewer per week.

6.3.1 Constructing a Frequency Table in R

There are many ways to build a frequency table in R. Here, we will show you two different ways. The first approach uses the dplyr package and gives us only the frequency (count) of categories. For the second approach, we will download a new package (questionr) and use it to build a frequency table similar to Table 6.1. Here is an example of how to build a frequency table using our two approaches:

Table 6.1 Frequency table of the variable TP (times partying per week)

On average, how many times per week do you party?	Frequency	Percentage	Cumulative %
0	3	7.5	7.5
1	8	20.0	27.5
2	10	25.0	52.5
3	9	22.5	75.0
4	8	20.0	95.0
5 or more	2	5.0	100.0
Total	40	100.0	

```
## Before building the table, we need to designate our vector as a factor.
factor(dat$TP)

## Approach 1: dplyr (only gives us the frequency of each category)
count(dat, TP)
  timespartying n
1             0 3
2             1 8
3             2 10
4             3 9
5             4 8
6             5 2

## Approach 2: questionr
install.packages("questionr")
library(questionr)
freq(dat$TP, cum = TRUE)
   n    %    val%  %cum   val%cum
0  3    7.5  7.5    7.5     7.5
1  8   20.0 20.0   27.5    27.5
2 10   25.0 25.0   52.5    52.5
3  9   22.5 22.5   75.0    75.0
4  8   20.0 20.0   95.0    95.0
5  2    5.0  5.0  100.0   100.0

## As we can see, the second approach gives us a frequency table with all the
information we need (frequency, %, and % cumulative). In order to obtain
cumulative %, we inserted the argument 'cum = TRUE' in the command.
```

6.4 Measures of Central Tendency

In this section, we shortly introduce the most widely used measures of central tendency (sometimes referred to as univariate statistics): the mean, the median, the mode, and the range.

6.4.1 Mean

The mean is the value we commonly call the average. To calculate the mean of a set of values, we must sum up all observations and then divide this number by the number of values or observations. In statistics, the mean is denoted by $\overline{x}$ for a sample mean and μ for a population mean (pronounced mu).

We can calculate a mean manually using the following formula:

$$\overline{x} = \frac{x_1 + x_2 + x_3 + \cdots + x_n}{n} = \frac{\sum_{i=1}^{n} x_i}{n},$$

where

$x_i =$ observations
$n =$ number of observations.

6.4.2 Median

The median, or "midpoint," is the middle number of a distribution. It is less sensitive to outliers and therefore a more "resistant" measure of central tendency than the mean. To calculate the median by hand, we just need to line up all values in order and find the middle one (or average of the two middle values when n, the number of observations, is even).

For instance, the following variable measures the grades of 11 students on an exam:

53 60 64 69 78 **79** 83 83 89 92 98

Here, the middle value of the distribution is 79. That is our median. In the event that there is an even number of values in a distribution, the media is the mean of the two central values.

6.4.3 Mode

The mode is the value that occurs most often in a given sample. In cases where there are several values that occur most often, the mode can consist of these several values (e.g., bimodal or multimodal distributions). Using the same distribution above, the value that appears most often is 83 (occurs twice), which is the mode of this subsample.

6.4.4 Range

The range is a measure that provides us with some indication of how widely spread out our data are. It is calculated by subtracting the maximum from the minimum (highest and lowest values in a given distribution). In the subsample used above, the range would be 45 (in this case, 98 minus 53).

6.4.5 Measures of Central Tendency in R

In R, we can compute measures of central tendency using base R commands. These include mean(), median(), and range(). It should be noted that there is no base command to measure the mode. However, we can rely on the count() function and identify the value which has the highest frequency. In the following example, we use our dataset on student's partying habits to show you how to compute measures of central tendency in R.

```
## Let's measure the mean, median, and range of the variable 'MSP'.
mean(dat$MSP)
[1] 76.5
median(dat$MSP)
[1] 70.
range(dat$MSP)
[1] 30 200
200 - 30
[1] 170

## Note how the range function does not calculate the range but rather gives
you the minimum value and maximum value. You can use these two values
to calculate the range by subtracting them (as shown above).
```

From the code above, we can conclude that the mean amount of money spent partying by student is 76.5$. The median is 70$, and our values range from 30$ to 200$ for a calculated range of 170$.

6.5 Displaying Data Graphically with Pie Charts, Boxplots, and Histograms

6.5.1 Pie Charts

A pie chart, sometimes also called a circle chart, is one way to display a frequency table graphically. The graphic consists of a circle that is divided into slices. Each slice represents one of the variable's categories. The size of each slice is proportional to the frequency of the value. Pie charts can be strong graphical representation of categorical data with few categories. However, the more categories we include into a pie chart, the harder it is to succinctly compare categories. It is also rather difficult to compare data across different pie charts.

Figure 6.2 displays the pie chart of the variable TP. We can see that the interpretation is already difficult, as it is already hard to guess the frequency of each slice from the graph. If we take another variable with more categories such as the variable money spent partying (see Fig. 6.3), we see that the categories are basically indistinguishable from each other. Hence, a pie chart is not a good option to display this variable.

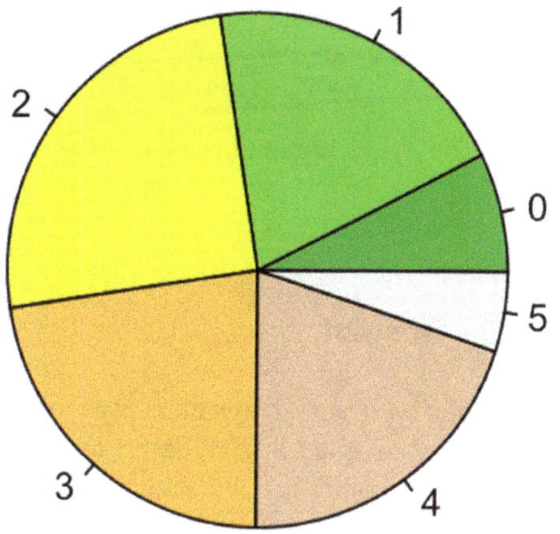

Fig. 6.2 Pie chart of the variable times partying (TP)

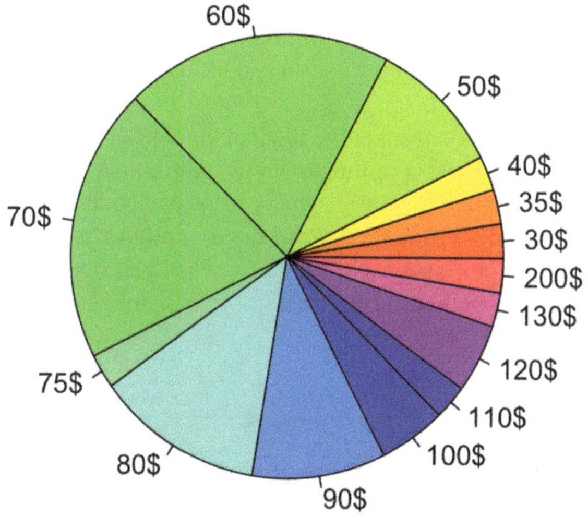

Fig. 6.3 Pie chart of the variable money spent partying (MSP)

To build a pie chart in R, we must first specify the size of the slices as well as the labels to be used for each slice as vectors. Then, we can use these vectors to build a pie chart using the pie() command. The following code replicates Fig. 6.3.

```
## Let's replicate the pie chart in Fig. 6.3. We begin by creating vectors with
the frequency of each of the six categories and creating a separate vector with
the label of each category.
slices <- c(1, 1, 1, 4, 8, 8, 1, 5, 4, 2, 1, 2, 1, 1)
labels <- c("30$","35$","40$","50$","60$","70$","75$","80$","90$",
            "100$","110$","120$", "130$","200$")
```

```
## Now we can use the pie() command to build the pie chart.
pie(slices, labels = labs, col = rainbow(14))

## The col = argument is used to specify the color palette of the pie chart.
```

6.5.2 Boxplot

A boxplot is a very convenient way of displaying variables. This type of graph allows us to display three measures of central tendency in one graph. The median (or the mid-point) is indicated by the black centerline in the box. The shaded box, which is also known as the inter-quartile range (IQR), includes the middle 50% of the data. The two outer lines denote the range of the data (minimum and maximum). If values extend up to 1.5 times the inter-quartile range from the upper or lower boundary of the mid-50%, they are plotted individually as asterisks. These individually plotted values are considered outliers.

Figure 6.4 displays the boxplot of the variable MSP in our party dataset. From the graph, we see that the median amount of money spent partying for the students who participated in the survey is approximately 70$/week. We also learn that the mid-50% of the money spent is between 60$ and 80$. The maximum value denoted by the upper line is 130$ and the minimum value denoted by the lower line is 30$. The boxplot also indicates that we have an outlier (the dot at the top of the plot). In R, we build a boxplot using the boxplot() command.

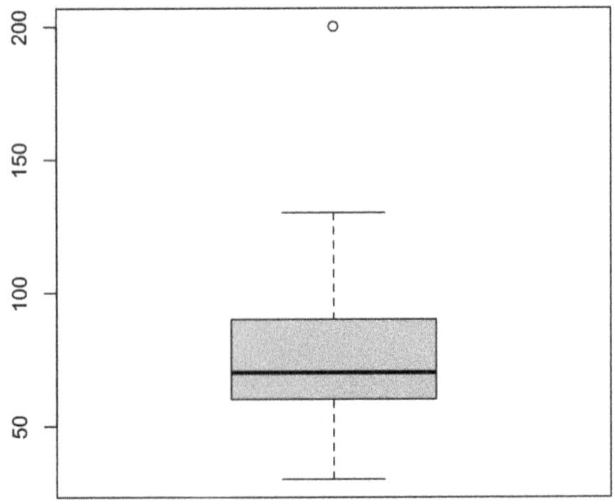

Fig. 6.4 Boxplot of variable money spent partying (MSP)

```
## Let's build a boxplot of the money spent partying variable using R
boxplot(dat$MSP)
```

6.5.3 Histogram

Histograms are one of the most widely used graphs in statistics to graphically display continuous variables. These graphs display the frequency distribution of a given variable. Histograms are very important for statistics in that they tell us about the distribution of the data. In statistical inference (i.e., when using a sample to generalize about a population), normally distributed data is a prerequisite for many statistical tests (see below) which we use to generalize from a sample toward a population.

Figure 6.5 illustrates two normal distributions (i.e., the blue line and the red line). In their ideal form, these distributions have the following features:

(1) The mode, mean, and median are the same value in the center of the distribution.
(2) The distribution is symmetrical, that is it has the same shape on each side of the distribution.

To build a histogram in R, we can rely on the base R function hist(). Figure 6.6 presents the histogram for the variable MSP in the party data. We can see from the distribution that the data is somewhat normally distributed with an outlier. We use the argument xlab= to specify the label for the x axis. We can also specify

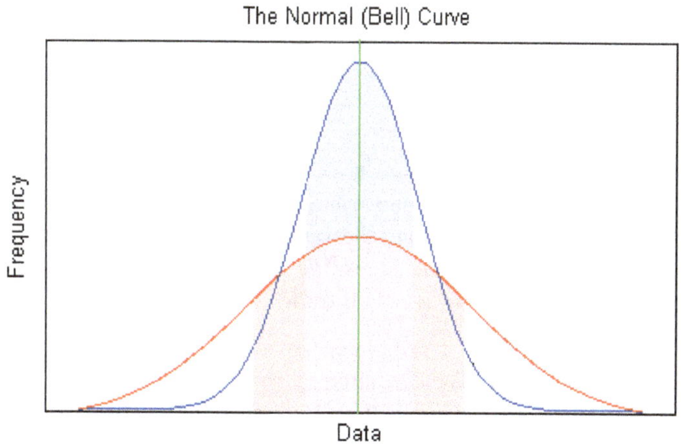

Fig. 6.5 Shape of a normal distribution

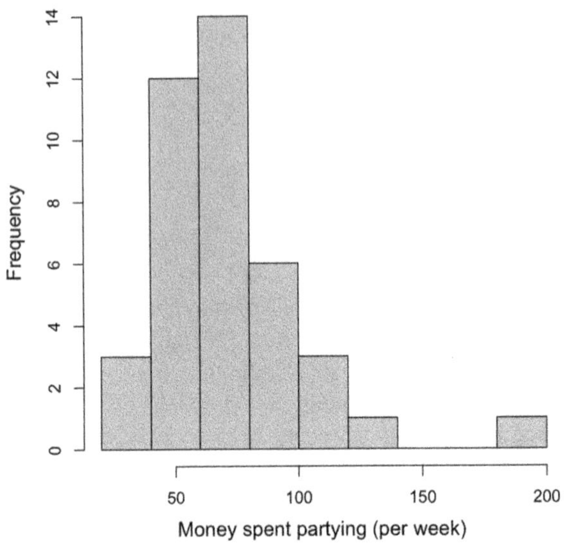

Fig. 6.6 Histogram of variable money spent partying (MSP)

a title for the histogram using the additional argument main=. When specifying a title or label, it is important to place quotation marks around the desired text to make sure R understands you want it to print text.

```
## Let's build a histogram of the variable MSP using the hist() command.
hist(dat$MSP, xlab = "Money spent partying (per week)")
```

6.6 Measures of Dispersion, Sampling Error, and Confidence Intervals

On the following pages, we illustrate how you can calculate the sampling error and the confidence interval, two univariate statistics that are of high value for survey research. In order to calculate the sampling error and confidence interval, we must follow several intermediate steps. We must calculate the deviation, sample variance, standard deviation and standard error.

Deviation. Every sample has a sample mean and for each observation there is a deviation from that mean. The deviation is positive when the observation falls above the mean and negative when the observation falls below the mean. The magnitude of the value reports how different (in the relevant numerical scale) an observation is from the mean.

$$deviation_i = x_i - \overline{x},$$

where

$x_i =$ observations
$\overline{x} =$ mean.

Sample Variance. The variance is the approximate average of the squared deviations. In other words, the variance measures the approximate average of the squared distance between observations and the mean. For this measure, we use squares because the deviations can be negative, and squaring gets rid of the negative sign. It should be noted that the variance can be symbolized using s^2 or σ^2 (sigma-squared). Usually, the latter is used to refer to the population variance while the former is used for a sample variance.

$$s^2 = \frac{\sum(x_i - \overline{x})^2}{n - 1},$$

where

$x_i =$ observations
$\overline{x} =$ mean
$n =$ number of observations
$s^2 =$ variance.

Standard Deviation. The standard deviation is a measure of volatility that measures the amount of variability or volatility around the mean. The standard deviation is large if there is high volatility in the data and low if the data is closely clustered around the mean. In other words, the smaller the standard deviation, the less "error" we have in our data and the more secure we can be in knowing that our sample mean closely matches our population mean.

$$s = \sqrt{\frac{\sum_{i=1}^{N}(x_i - \overline{x})^2}{n - 1}},$$

where

$x_i =$ observations
$\overline{x} =$ mean
$n =$ number of observations
$s =$ standard deviation.

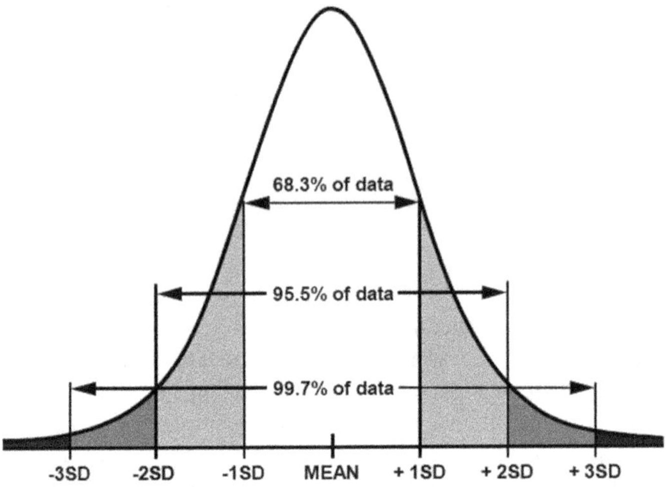

Fig. 6.7 Standard deviation in a normal distribution

The standard deviation is also important for standardizing variables. If the data is normally distributed (i.e., they follow a bell-shaped curve), the data has the following properties. 68.3% of the cases fall within one standard deviation of the mean, 95.5% of the cases fall between two standard deviations from the mean, and 99.7 cases fall between three standard deviations from the mean (see Fig. 6.7). It should be noted that the standard deviation can be symbolized using s or σ (sigma). Usually, the latter is used to refer to the population standard deviation.

Standard Error. The standard error allows researchers to measure how close the mean of a given sample is to the population mean. It is calculated by dividing the standard deviation of the sample by the root of the sample size (n).

$$SE = \frac{s}{\sqrt{n}},$$

where

$SE =$ standard error
$n =$ number of observations
$s =$ standard deviation.

The standard error is important because it allows researchers to calculate the confidence interval. The confidence interval, in turn, allows researchers to make inferences from the sample mean toward the population mean. It also allows researchers to calculate the population mean based on the sample mean. In other words, it gives us a range where the real mean falls. However, this method only works if we have a random sample and a normally distributed variable.

Confidence Interval. Surveys generally use the confidence interval to depict the accuracy of their predictions. We can calculate the confidence interval at different levels, with the most common being 90%, 95%, and 99% (usually, by default, we use the 95% confidence interval)

$$CI = \overline{x} \pm z * \left(\frac{s}{\sqrt{n}}\right),$$

where

$CI =$ confidence interval
$n =$ number of observations
$s =$ standard deviation
$z =$ Z-score corresponding to the confidence level (1.96 for 95% confidence).

For example, a 2006 opinion poll of 1000 randomly selected Americans aged 18 to 24 conducted by the Roper Public Affairs Division and National Geographic finds that:

- 63% of young adults ages 18–24 cannot find Iraq on a map of the Middle East.
- 88% young adults ages 18–24 cannot find Afghanistan on a map of Asia.

At the end of the survey, we find the stipulation that the results of this survey are accurate at the 95% confidence level plus or minus 3 percentage points (Margin of error ± 3%). This means that we are 95% confident that the *true population* statistic (i.e., the true percentage of American youths, in the whole population, who cannot find Iraq on a map) is somewhere in between 60 and 66%. In other words, the "real" mean in the population is anywhere between plus and minus 3 percentage points from the sample mean. This error range is normally called the margin of error or the **sampling error**. In the Iraqi example, we can say that we have a sampling error of ±3 percentage points from the sample mean (see Fig. 6.8).

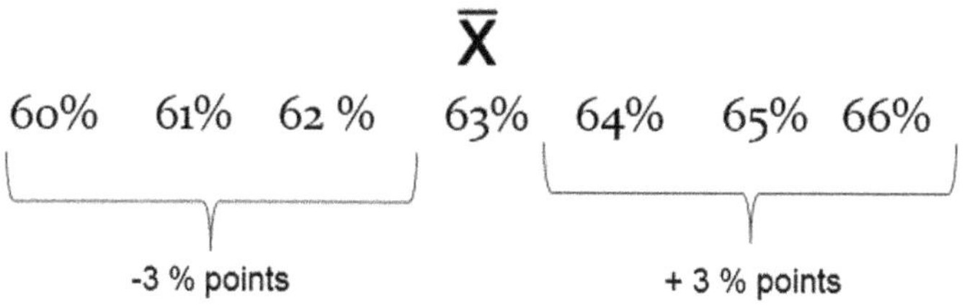

Fig. 6.8 Graphical depiction of the confidence interval

6.6.1 Calculating Confidence Intervals in R

With the help of R, we can calculate the standard deviation and the standard error. We cannot directly calculate the confidence interval, but instead we rely on R to calculate it using the formula above (i.e., we have to do the last step by hand). To do so, we must first begin by calculating the standard deviation and the standard error. The following example calculates the confidence interval for the variable of ST (study time per week) in our dataset on party habits of students. In this case, we will use a 95% CI (confidence interval).

```
## Let's begin by retrieving the mean of the data.
mean(dat$ST)
[1] 9.375

## Now, let's find the standard deviation of the data.
StDev <- sd(dat$ST)
StDev
[1] 3.094143

## Now, let's calculate the standard error.
SE <- StDev / sqrt(40)
SE
[1] 0.2823891

## Finally, we can calculate the confidence interval using z-value of 1.96.
CI <- 1.96 * SE
CI
[1] 0.9588847

## We can conclude that our 95% confidence interval is our mean + or −
our confidence interval. In substantive terms, this means that, if this was a
random or fully representative sample of the US population, we could be
95% confident that undergraduate students (in the whole population) go out
to party 2.3 times per week, on average, with a margin of error of±0.55.
In other words, we would be 95% confident that the average undergraduate
student parties somewhere between 1.75 and 2.85 times per week.
```

References

Arel-Bundock, V. (2022). modelsummary: Data and model summaries in R. *Journal of Statistical Software, 103*(1), 1–23. https://doi.org/10.18637/jss.v103.i01

Wickham, H., Averick, M., Bryan, J., Chang, W., McGowan, L. D., François, R., Grolemund, G., Hayes, A., Henry, L., Hester, J., Kuhn, M., Pedersen, T. L., Miller, E., Bache, S. M., Müller, K., Ooms, J., Robinson, D., Seidel, D. P., Spinu, V., … Yutani, H. (2019). Welcome to the tidyverse. *Journal of Open Source Software, 4*(43), 1686. https://doi.org/10.21105/joss.01686

Further Readings and Resources

R Introductory Books

Gillespie, B. J., Hibbert, K. C., & Wagner III, W. E. (2021). *A guide to R for the social and behavioral science statistics*. Sage Publishing.

A great reference book for learning how to use R and RStudio. The book covers most introductory topics needed to begin using R as a social scientist.

Wickham, H., & Grolemund, G. (2016). *R for data science: Import, tidy, transform, visualize, and model data*. O'Reilly Media Inc.

This book presents all the necessary tools to use R for data science. It is a tremendous resource when it comes to cleaning and transforming data as well as for basic modeling needs.

Arel-Bundock, V. (2020). *Analyse causale et méthodes quantitatives: une introduction avec R, Stata et SPSS*. Presses de l'Université de Montréal.

This book is a great resource for French speakers. It introduces all the necessary mathematical and statistical knowledge you need to begin working with R.

R Web Resources

https://www.r-project.org—Official website for the R project. Useful for information about the R programming language and updates regarding the project.

https://www.statmethods.net/r-tutorial/index.html—A great web reference for all things R syntax. If you have questions about how to write a line of code and what various operators mean, this is the place to go.

https://stackoverflow.com—Stackoverflow is an online community where users can post questions regarding their R code. Oftentimes, the problem you are having with your code or models has already been someone else's problem. You can usually find answers to your R questions on this platform.

https://www.statology.org—Statology is a website which includes several tutorials for basic R and statistics. We recommend looking through the in-depth explanations which often include great examples.

Univariate and Descriptive Statistics

Park, H. M. (2008). *Univariate Analysis and Normality Test Using SAS, Stata, and SPSS. Technical Working Paper*. The University Information Technology Services (UITS) Center for Statistical and Mathematical Computing, Indiana University. Accessible at: https://scholarworks.iu.edu/dspace/handle/2022/19742.

A concise introduction into descriptive statistics and graphical representations of data including a discussion of their underlying statistical assumptions.

Bivariate Statistics with Categorical Variables

7

7.1 Independent Samples t-Test

An independent samples t-test assesses whether the means of two groups are *statistically* different from each other. To properly conduct such a test, the following conditions should be met:

1. The dependent variable should be continuous.
2. The independent variable should consist of mutually exclusive groups (i.e., be categorical).
3. All observations should be independent, which means that there should not be any linkage between observations (i.e., there should be no direct influence from one value within one group over other values in this same group).
4. There should not be many significant outliers (especially with smaller samples).
5. The dependent variable should be more or less normally distributed.
6. The variances between groups should be similar.

For example, for our data we might be interested whether male respondents spend more money than female respondents while partying. In this case, our dependent variable would be money spent partying (per week) and our independent variable gender. We have relative independence of observations as we cannot assume that the money one individual in the sample spends partying directly hinges upon the money another individual in the sample spends partying. From Fig. 6.6, we also know that the variable money spent partying is approximately normally distributed. As a preliminary test, we must check if the variance of our two distributions is similar.

Having verified that our data fit the conditions for a t-test, we can now get into the mechanics of conducting such a test. Intuitively, we could first compare the means for the two groups. In other words, we should look at how far the two means are apart from each other. Second, we ought to look at the variability of

© The Author(s), under exclusive license to Springer Nature Switzerland AG 2023

D. Stockemer and J.-N. Bordeleau, *Quantitative Methods for the Social Sciences*, Springer Texts in Political Science and International Relations,
https://doi.org/10.1007/978-3-031-34583-8_7

the data. Pertaining to the variability, we can follow a simple rule; the less there is variability, the less there is overlap in the data, and the more the two groups are distinct. Therefore, to determine whether there is a difference between two groups, two conditions must be met: (1) the two group means must differ quite considerably and (2) the spread of the two distributions must be relatively low. More precisely, we must judge the difference between the two means relative to the spread or variability of their scores (see Fig. 7.1). The t-test does just this.

Figure 7.2 graphically illustrates that it is not enough that two group means are different from one another. Rather, it is also important how closely the values of the two groups cluster around a mean. In Case I, we can see that the two groups are distinct (i.e., there is little data overlap between the two groups). In Case II, we can be rather sure that these two groups are similar (i.e., more than 80% of the data points are indistinguishable; they could belong to either of the two groups). These graphs show the importance of variability when measuring the difference between two groups (i.e., t-test).

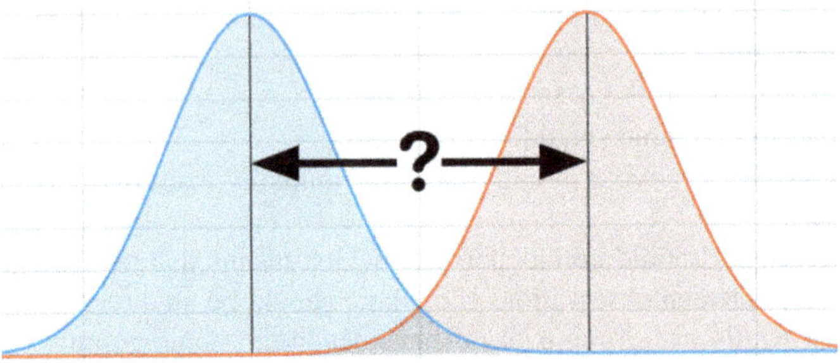

Fig. 7.1 The logic of a t-Test

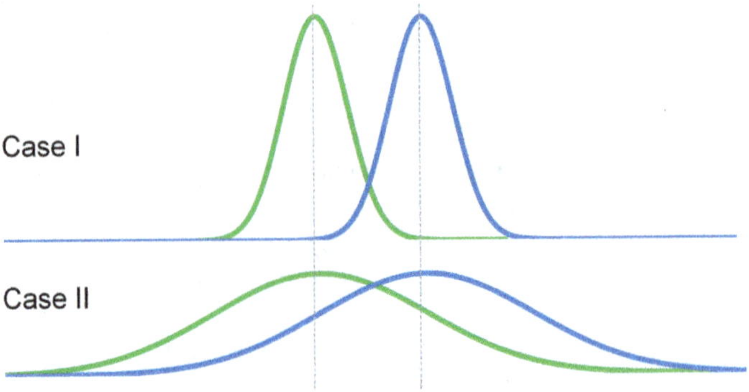

Fig. 7.2 The role of variability in a t-Test

7.1.1 Calculating a t-Value for Independent Samples t-Test

The difference between the means is the signal and the bottom part of the formula is the noise, or a measure of variability; the smaller there are differences in the signal and the larger the variability, the harder it is to see group differences. The logic of a t-test can be summarized as follows:

$$t = \frac{\text{signal}}{\text{noise}} = \frac{\text{difference between group means}}{\text{variability of groups}}$$

$$t = \frac{\overline{x}_T - \overline{x}_C}{\text{SE}(\overline{x}_T - \overline{x}_c)}$$

The top part of the formula is easy to compute—just find the difference between the means. The bottom is a bit more complex; it is called the **standard error of the difference**. To compute it, we have to take the variance for each group and divide it by the number of people in that group. We add these two values and then take their square root. The specific formula is as follows:

$$\text{SE}(\overline{x}_T - \overline{x}_c) = \sqrt{\frac{\text{var}_T}{n_T} + \frac{\text{var}_C}{n_C}}$$

The final formula for the t-value is the following:

$$t = \frac{\overline{x}_T - \overline{x}_C}{\sqrt{\frac{\text{var}_T}{n_T} + \frac{\text{var}_C}{n_C}}}$$

The t-value will be positive if the first mean is larger than the second one is and negative if it is smaller. However, for our analysis this does not matter. What matters more is the size of the t-value. Intuitively, we can say that the larger the t-value the higher the chance that two groups are statistically different. A high t-value is triggered by a considerable difference between the two group means and low variability of the data around the two group means. To statistically determine whether the t-value is large enough to conclude that the two groups are statistically different we must use a test of significance. A test of significance sets the amount of error, called the alpha level, which we allow our statistical calculation to have. In most social research, the 'rule of thumb' is to set the alpha level at 0.05. This means that we allow five percent error. In other words, we want to be 95% certain that a given relationship exists. This implies that, if we were to take 100 samples from the same population, we could get a significant t-value in 95 out of 100 cases.

As you can see from the formula, doing a t-test by hand can be rather complex. Therefore, we turn to R to help us.

7.1.2 Doing an Independent Samples t-Test in R

Step 1. Pre-test—Create a histogram to detect whether the dependent variable—Money Spent Partying—is normally distributed (see the histogram we built in Sect. 6.5.3. and use the same code). Despite the outlier ($200 per week), the data is approximately normally distributed, and we can proceed with the independent samples t-test (see Fig. 7.3).

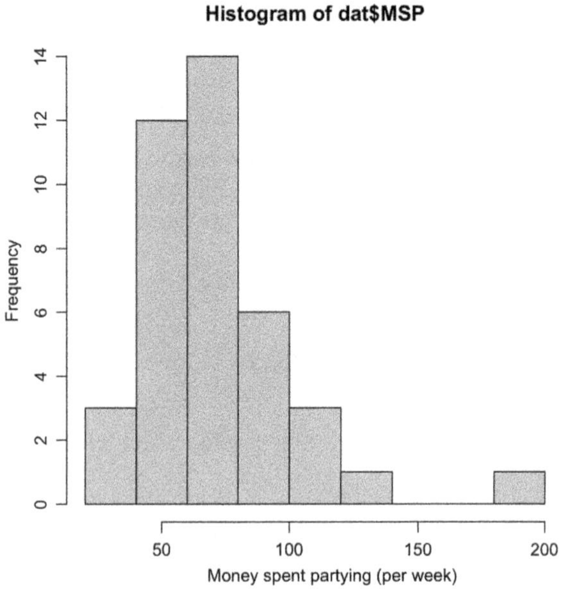

Fig. 7.3 Histogram of Variable MSP (money spent partying)

Step 2. Verifying equal variance assumption—Before we conduct and interpret the t-test, we have to verify whether the assumption of equal variance is met. To do so, we conduct an F-test (also known as Levene's test). If the F-value is not significant (i.e., the significance level in the second column is larger than 0.05), we do not violate the assumption of equal variances. We use the command var.test() to conduct an F-test in R.

```
## Let's conduct the F-test to see if the variance is the same for the variable
MSP in both gender groups.
```

```
var.test(MSP ~ Gender, data = dat, alternative = "two.sided")

## Here is the output for the F-test.

            F test to compare two variances

data:  MSP by Gender
F = 0.15927, num df = 18, denom df = 20, p-value = 0.0002549
alternative hypothesis: true ratio of variances is not equal to 1
95 Percent Confidence Interval:
0.06367125 0.40756812
Sample estimates:
ratio of variances
        0.1592683
```

In this case, the significance value is below 0.05 ($p = 0.0002$). This implies that the assumption of equal variances is violated. Yet, this is not dramatic for interpreting the t-test since we have an outlier value which skews the variance of one of our groups (male group).

Step 3. Compute t-test—Use the t.test() command to compute a t-test using Money Spent Partying as the dependent variable and gender as the group categories. As we can see, the t.test() command includes a number of arguments. The "MSP ~ Gender" argument stipulates the dependent variable (MSP) and the variable for the two groups (Gender). The "var.equal = FALSE" argument stipulates that we have not met the equality of variance assumption. Lastly, argument "data = dat" specifies what data frame the data should come from (in this case, the data frame is the one we imported and named 'dat').

```
## Let's conduct a t-test for the variable MSP based on gender.
t.test(MSP ~ Gender, var.equal = FALSE, data = dat)

## Results of the t-test computed in R

                Welch Two Sample t-test

data: MSP by Gender
t = -0.63239, df = 26.74, p-value = 0.5325
alternative hypothesis: true difference in means between group 0 and
group 1 is not equal to 0
95 Percent Confidence Interval:
-24.90147  13.17215
```

```
sample estimates:
mean in group 0 mean in group 1
       73.42105         79.28571
```

As we can see, the t.test() command includes a number of arguments. The "MSP ~ Gender" argument stipulates the dependent variable (MSP) and the variable for the two groups (Gender). The "var.equal = FALSE" argument stipulates that we have not met the equality of variance assumption. Lastly, argument "data = dat" specifies what data frame the data should come from (in this case, the data frame is the one we imported and named 'dat').

7.1.3 Interpreting an Independent Samples t-test

Having tested the data for normality and equal variances, we can now interpret the t-test. We find that female respondents (who were coded 1), spend slightly more money when they go out and party compared to male respondents (which we coded 0). Yet, the difference is rather moderate. On average, female participants merely spend 6 dollars more per week than male participants.

As we have discussed, mean difference is not the only factor to consider when conducting a t-test. The variance in each group (and the overlap of such variance) is also important. The t-test helps us by displaying the significance or alpha level of the t-value. Assuming that we take the 0.05 benchmark, we cannot reject the null hypothesis with 95% certainty because our p-value is 0.53 (larger than our benchmark). Hence, we can conclude that that there is no statistically significant difference between the two groups. In other words, there is no significant difference in the amount of money spent partying per week between male and female respondents.

7.1.4 Reporting the Results of Our Independent Samples t-test

In Sect. 4.11.1, we hypothesized that male students like the bar scene more than female students do and are therefore going to spend more money when they go out and party. The independent samples t-test disconfirms this hypothesis. On average, it is actually female students who spend slightly more. However, the difference in average spending (73 dollars for guys and 79 dollars for girls) is not statistically different from zero ($p = 0.53$). Hence, we cannot reject the null-hypothesis and can conclude that the spending pattern for partying are similar for the two genders.

7.2 One-Way Analysis of Variance (ANOVA)

T-tests work great with dichotomous categorical variables (when we have two groups), but sometimes we have categorical variables with more than two categories. In cases, where we have a continuous variable paired with an ordinal or nominal variable with more than two categories, we use what is called a one-way ANOVA (or F-test). The logic behind an ANOVA is similar to the logic for a t-test. If we consider the graph in Fig. 7.4, we can see that some groups are different while others are not. Groups A and D are completely different while groups C and D are not. In order to test if groups A, B, C, and D are statistically different from one another, we must conduct a one-way analysis of variance.

While the logic of an ANOVA reflects the logic of a t-test, the calculation of several group means and several measures of variability around the group means becomes more complex in an F-test. To reduce this complexity, an ANOVA test uses a simple method to determine whether there is a difference between several groups. It splits the total variance into two groups: between variance and within variance. The between variance measures the variation between groups, whereas the within variance measures the variation within each group. Whenever the between variation is considerably larger than the within variation, we can say that there are differences between groups. The following example highlights this logic (see Fig. 7.5).

Let us assume that Fig. 16 depicts two hypothetical samples which measure the approval ratings of German Chancellor Scholz based on socioeconomic class. In the survey, an approval score of 0 means that respondents are not at all satisfied with his performance as chancellor. In contrast, 100 signifies that individuals are very satisfied with hid performance as chancellor. The first sample consists of young people (i.e., 18–25) and the second sample of old people (i.e., 65 and older). Both samples are split into three categories—high, medium and low. High stands for higher or upper classes, med stand for medium or middle classes, and low stand for the lower or working classes. We can see that the mean satisfaction ratings for

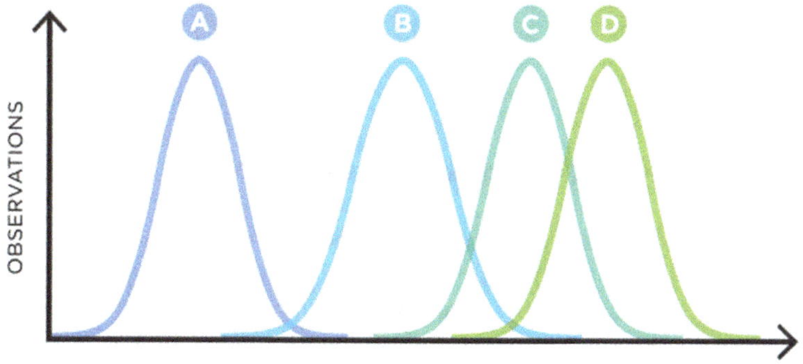

Fig. 7.4 What makes groups different?

	Sample A			Sample B		
	High	Med	Low	High	Med	Low
	50	40	30	30	28	12
	51	41	31	40	32	18
	52	42	32	55	40	30
	53	43	33	65	50	45
	54	44	34	70	60	55
Mean	**52**	**42**	**32**	**52**	**42**	**32**

Fig. 7.5 Within- and between-group variation

Chancellor Scholz per social strata do not differ between the two samples; that is, the higher classes, on average, rate him at 52, the middle classes at 42, and the lower classes at 32. However, what differs tremendously between the two samples is the variability of the data.

In the first sample, the values are very closely clustered around the mean throughout each of the three categories. We can see that there is much more variability between groups than between observations within one group. Hence, we would conclude that the groups are different. In contrast, in sample two, there is large within-group variation. That is, the values within each group differ much more than the corresponding values between groups. Therefore, we would predict for sample two that the three groups are probably not that different, despite the fact that their means are different. Following this logic, the formula for an ANOVA is **between group variance/within group variance**. Since it is too difficult to calculate the between and within group variance by hand, we let statistical computer programs do it for us.

A standard ANOVA test illustrates if there are differences between groups or group means, but it does not show which specific group means are different from one another. Yet, in most cases researchers want to know not only that there are some differences, but also between which groups the differences lie. So-called multiple comparison tests—basically t-tests between the different groups—compare all means against one another and help us detect where the differences lie.

7.2.1 One-Way Analysis of Variance in R

For our ANOVA test, we use the variable money spent partying per week as the dependent variable and the categorical variable times partying per week as the factor or grouping variable. Given that we only have 40 observations and given that there should be at least several observations per category to yield valid test results, we reduce the six categories to three. In more detail, we cluster together no partying and partying once; partying twice and three times; and partying four times and five time and more together to create three categories. We can create this new variable in R, which we will label TP1.

```
## Let's begin by creating the new variable TP1 from existing variable TP.
dat$TP1 <- recode(
  dat$TP, '0' = 1,'1' = 1,'2' = 2,'3' = 2,'4' = 3,'5' = 3)

## Now let us conduct the ANOVA test (F-test) with TP1

oneway.test(MSP ~ TP1, data = dat, var.equal = TRUE)

## As we can see, the arguments for the oneway.test() command are the same
as for the t.test() command. Here are the results of our one-way ANOVA in
R:
            One-way analysis of means

data: MSP and TP1
F = 10.491, num df = 2, denom df = 37, p-value = 0.0002459
```

7.2.2 Interpreting the Results of an ANOVA

The ANOVA results in R provide us with an F-score and a significance level. Similar to a t-test, the most important result is the significance level (i.e., p-value). It tells us whether there is a difference between at least two of the groups we have. In our example, the significance level is $p = 0.0002$, which means that we can tell with nearly 100% certainty that at least two groups differ in the money they spent partying per week. With that being said, these results do not tell us which groups are different; the only thing they tell us is that at least two groups are different from one another. The next section addresses this problem by conducting post-hoc tests.

7.2.3 Post-hoc or Multiple Comparison Tests in R

The violation of the equal variance assumption (i.e., the distributions around the group means are different for the various groups in the sample) is rather unproblematic for interpreting a one-way ANOVA (Park, 2009). However, having an equal or unequal variability around the group means is important when doing multiple pairwise comparison tests between means. We must therefore test this assumption before we can do the post-hoc comparison tests. In R, we use the Bartlett's test of homogeneity of variances to see if the variance between three or more groups are different or not.

```
## Let's conduct the Bartlett's to see if the variance is the same for the variable MSP across the three groups.
bartlett.test(MSP ~ TP1, data = dat)

## Here is the output for the Bartlett's test.

        Bartlett test of homogeneity of variances

data: MSP by TP1
Bartlett's K-squared = 17.252, df = 2, p-value = 0.0001793
```

The homogeneity of variances test provides a significant result (p-value = 0.0001) indicating that the variation around the three group means in our sample is not equal. We have to take this inequality of variance in the three distributions into consideration when conducting the multiple comparison test.

There are different types of post-hoc comparison tests. For instance, when we want to compare all groups to each other and we meet the equality of variance assumption, we can use the Tukey HSD test. When we have a reference group (i.e., a control group), we can use the Dunnett test. If we already have a planned comparison between specific groups, we can use the Bonferroni correction test. In our case, since we want to compare all groups and have not met the equality of variance assumption, we will conduct Tamhane's T2 test.

```
## Computing Tamhane's T2 all-pairs comparison test for unequal variances. Notice that we must first download the package PMCMRplus.
install.packages("PMCMRplus")
library(PMCMRplus)
tamhaneT2Test(MSP ~ TP1, data = dat)
## Here is the output for the Tamhane's T2 test.
        Pairwise comparisons using Tamhane's T2-test for unequal
variances
data: MSP by TP1

  1   2
2 0.649 -
```

```
3 0.022 0.051

P value adjustment method: T2 (Sidak)
alternative hypothesis: two.sided
```

The post hoc multiple comparison test allows us to decipher which groups are actually statistically different from one another. In R, the test gives us the p-value for the comparison tests between the various groups. From the output of the test, we can see that individuals in group 3 are statistically different from those in group 1 ($p = 0.022$). This is the only statistically significant difference in our three groups, although group 2 and 3 are almost significantly different ($p = 0.051$). Therefore, those who party 4 or more times per week spend significantly more than those who party 1 time or less. These are the only two groups that share a statistically significant difference.

7.2.4 Reporting the Results of an ANOVA and Post-hoc Comparison Tests

In Sect. 4.11.1, we hypothesized that individuals who party more frequently will spend more money for their weekly partying habits than individuals that party less frequently. Creating three groups of party goers—(1) students who party once or less; (2) students who party between two and three times; and (3) students who party more than four times—we find some support for our hypothesis. Our one-way ANOVA indicates that there are differences between groups ($p = 0.0002$). Yet, the F-test cannot tell us between which groups the differences lie. In order to find this out, we computed a post-hoc multiple comparison test. We do so assuming unequal variances between the three distributions, because a Bartlett's test of equal variance ($p = 0.0001$) reveals that the null hypotheses of equal variances must be rejected. Our results indicate that the mean spending average statistically significantly differs between groups 1 and 3 (i.e., the average difference in party spending is significant at $p = 0.022$), but not between other groups.

7.3 Cross-Tabulation Tables and Chi-Square Test

7.3.1 Cross-Tabulation Tables

So far, we have discussed bivariate tests that work with a categorical variable as the independent variable and a continuous variable as the dependent variable (i.e., an independent samples t-test if the independent variable is binary and the dependent variable continuous, and an ANOVA if the independent variable has more than two categories). What happens if both the independent and dependent variable are binary? In this case, we can present the data in a crosstab (cross-tabulation tables).

Table 7.1 Two-by-two table of the relationship between drug treatment and the survival status

	Dead	Alive
Treated	36	14
Not treated	30	25
Total	66	39

Table 7.1 provides an example of a two-by-two table. The table presents the results of a lab experiment with mice. A researcher has 105 mice with a severe illness; she treats 50 mice with a new drug and does not treat 55 mice at all. She wants to know whether this new drug can cure the illness. To do so, she creates four categories: (1) treated and dead, (2) treated and alive, (3) not treated and dead, and (4) not treated and alive.

Based on this two-by-two table we can ask the question: How many of the dead are either treated or not treated. To answer this question, we have to use the column as unit of analysis and calculate the percentage of dead which are treated and the percentage of dead which are not treated. To do so, we have the convert the column's raw numbers into percentages. To calculate these percentages for the first field, we take the number in the field—treated/dead—and divide it by the column total (36/66 = 55.55%). We do analogously for the other fields (see Table 7.2). Interpreting Table 7.2, we can find, for example, that roughly 56% of the dead mice have been treated, but only 35.9% of those alive have undergone treatment.

Instead of asking the question how many of the dead mice have been treated, we might change the question and ask how many treated mice are alive or dead? To get this information, we first calculate the percentage of dead mice with treatment and of alive mice with treatment. We do analogously for the dead that are not treated and the alive that are not treated. Doing so, we find that off all treated 72% are dead and only 28% are alive. In contrast, the non-treated mice have a higher chance of survival. Only 55.55% have died and 44.45% have survived (see Table 7.3).

Table 7.2 Two-by-two table focusing on the interpretation of the columns

	Dead	Alive
Treated	36	14
Percent	55.55%	35.90%
Not treated	30	25
Percent	44.45%	64.10%
Total	66	39
Percent	100%	100%

Table 7.3 Two by two table focusing on the interpretation of the rows

	Dead	Alive	Total
Treated	36	14	50
Percent	72.00%	28.00%	100%
Not treated	30	25	55
Total	55.55%	44.45%	100%

7.3.2 Chi-Square Test of Independence

Having interpreted the crosstabs, a researcher might ask whether there is a relationship between drug treatment and the survival of mice. To answer this research question, she might postulate the following null hypothesis (and alternate hypothesis).

H_0: The survival of the animals is independent of drug treatment.

H_a: The survival of the animals is higher with drug treatment.

In order to determine if there is a relationship between drug treatment and the survival of mice, we use what is called a chi-square test of independence. To apply this test, we compare the actual value in each field with a random distribution of values between the four fields. In other words, we compare the real value in each field with an expected value, calculated so that the chance that a mouse falls in each of the four fields is the same.

To calculate this expected value, we use the following formula:

$$\frac{\text{Row Total} * \text{Column Total}}{\text{Table Total}}$$

Table 7.4 displays the observed and the expected values for the four possible categories: (1) treated and dead, (2) treated and alive, (3) not-treated and dead, (4) not-treated and alive. The logic behind a chi-square test is that the larger the gap between one or several observed and expected values, the higher the chance that there actually is a pattern or relationship in the data (i.e., that treatment has an influence on survival).

The formula for a chi-square (χ^2) test is:

$$\chi^2 = \sum \frac{(\text{Observed} - \text{Expected})^2}{\text{Expected}}$$

In more detail the four steps to calculate a Chi-Square Value are the following:

Table 7.4 Calculating the expected values

	Dead	Alive	Total
Treated	36 **(31.4)**	14 **(18.6)**	50
Not Treated	30 **(34.6)**	25 **(20.4)**	55
Total	66	39	105

(1) For each observed value in the table, subtract the corresponding expected value (O − E)
(2) Square the difference $(O - E)^2$
(3) Divide the squares obtained for each cell in the table by the expected number for that cell $(O - E)^2/E$.
(4) Sum all the values for $(O - E)^2/E$. This is the chi-square statistic.

Our treated animal example gives us a chi-square value of 3.418. This value is not high enough to reject the null hypothesis which stipulates independence between the treatment and survival status.

A chi-square test works with the following limitations:

- No expected cell count can be less than 5.
- Larger samples are more likely to trigger statistically significant results.
- The test only identifies *that* a difference exists, not necessarily *where* it exists (If we want to decipher where the differences are, we have look at the data and detect in what cells are the largest differences between observed and expected values).

7.3.3 Chi-Square Tests in R

The dependent variable of our study which we used for the previous tests (i.e., t-test and ANOVA)—money spent partying—is continuous and therefore we cannot use it for a chi-square test. We have to use two categorical variables and make sure that there are least 5 observations in each cell. From our dataset, we can use the two categorical variables of gender and times partying per week. Since we have only 40 observations, we further contract the variable times partying and create two categories: partying a lot (three times or more a week) and partying moderately (less than 3 times a week). We name this new variable TP2. To create this new variable, see the previous example on recoding a variable in Sect. 7.2.1.

```
## Let's compute a chi-square test of independence for the variables Gender
and TP2 (which we recoded from TP) using the command chisq.test().
chisq.test(dat$TP2, dat$Gender)

## Here is the output for the Chi-Square Test:

        Pearson's Chi-squared test with Yates' continuity correction
```

```
data:  dat$TP2 and dat$Gender
X-squared = 0.090703, df = 1, p-value = 0.7633
```

7.3.4 Interpreting a Chi-Square Test Conducted in R

The results of our chi-square test reveal that we fail to reject the null hypothesis. There is therefore no relationship between gender and times partying. Our test value is 0.09 and the significance level is $p = 0.76$. Since our *p*-value is higher than our benchmark of 0.05, we can conclude that there are two groups (in this case gender) are not dependent when it comes to the average number of times they party per week. Based on these test-statistics, we indeed cannot reject the null hypothesis; this implies that we can conclude that the partying habits of male and female students (in terms of the times per week both genders party) do not differ.

7.3.5 Reporting the Results of a Chi-Square Test

Using a chi square test, we have tried to detect if there is a relationship between gender and the times per week students go out and party. We find that in absolute terms, roughly about half the female and male students, respectively, either party two times or less or three times or more. The chi-square test confirms that there is no statistically significant difference in the number of times either of the two groups go out to party (i.e., the chi square value is 0.09 and the corresponding significance level is 0.76).

References

Park, H. (2009). *Comparing group means: T-tests and one-way ANOVA using Stata, SAS, R, and SPSS*. Working Paper. The University Information Technology Services (UITS) Center for Statistical and Mathematical Computing, Indiana University.

Further Reading

Statistics Textbooks

Basically every introductory book to statistics covers bivariate statistics between categorical and continuous variables. The books we list here are just a short selection of possible textbooks. We have chosen these books because they are accessible and approachable and they do not use math excessively.

Brians, C. L. (2016). *Empirical political analysis: Pearson new international edition coursesmart etextbook*. Routledge (Chap. 11).

Provides a concise introduction into different types of means testing.
Macfie, B. P., & Nufrio, P. M. (2017). *Applied statistics for public policy*. Routledge.
This practical text provides students with the statistical tools needed to analyze data. It also shows through several examples how statistics can be used as a tool in making informed, intelligent policy decisions (part 2).
Morgan, S., Reichert, T., & Harrison, T. R. (2016). *From numbers to words: Reporting statistical results for the social sciences*. Routledge.
This book complements statistics books by showing scholars how they can present their test results in either visual or text form in an article or scholarly book. For students, this can also be useful for presenting test results in research papers.
Walsh, A., & Ollenburger, J. C. (2001). *Essential statistics for the social and behavioral sciences: A conceptual approach*. Prentice Hall (Chaps. 7–11).
These chapters explain in rather simple forms the logic behind different types of statistical tests between categorical variables and provide real life examples.

Bivariate Statistics with Two Continuous Variables

8

8.1 What is a Bivariate Relationship Between Two Continuous Variables?

A bivariate relationship involving two continuous variables can be displayed graphically and through a correlation or regression analysis. Such a relationship can exist if there is a *general* tendency for these two variables to be related, even if it is not in a completely determined way. In statistical terms, we say that these two variables "vary together"; this means that values of the variable x (the independent variable) tend to occur more often with some values of the variable y (the dependent variable) than with other values of the variable y.

8.1.1 Positive and Negative Relationships

When we describe relationships between variables, we normally distinguish between positive and negative relationships.

Positive relationship: high, or above average, values of x tend to occur with high, or above average, values of y. Also, low values of x tend to occur with low values of y.

Examples:

- Income and education
- National wealth and degree of democracy
- Height and weight.

Negative relationship: high, or above average, values of x tend to occur with *low*, or *below* average, values of y. Also, low values of x tend to occur with high values of y.

© The Author(s), under exclusive license to Springer Nature Switzerland AG 2023

D. Stockemer and J.-N. Bordeleau, *Quantitative Methods for the Social Sciences*, Springer Texts in Political Science and International Relations,
https://doi.org/10.1007/978-3-031-34583-8_8

Examples:

- State social spending and income inequality
- Exposure to Fox News and support for Democrats
- Smoking and life expectancy.

8.2 Scatterplot

A scatterplot graphically represents a quantitative relationship between two continuous variables. Each dot (point) is one individual observation's value on X and Y. The values of the independent variable (X) appear in sequence on the horizontal or x-axis. The values of the dependent variable (Y) appear on the vertical or y-axis. For a positive association, the points tend to move diagonally from lower left to upper right. For a negative association, the points tend to move from upper left to lower right. For NO association, points are scattered with no discernable direction.

8.3 Positive Relationship Displayed in a Scatterplot

Figure 8.1 displays a positive association, or a positive relationship, between countries' per capita GDP and the amount of energy they consume. We see that even if the data do not exactly follow a line, there is nevertheless a tendency that countries with higher GDP per capita values are associated with more energy usage. In other words, higher values of the x-axis (the independent variable) correspond to higher values on the y-axis (the dependent variable).

8.4 Negative Relationship Displayed in a Scatterplot

Figure 8.2 displays a negative relationship between per capita GDP and the share agriculture which makes up of a country's GDP. The figure clearly depicts a negative slope; that is, our results indicate that the richer a country becomes the more agriculture loses its importance for the economy. In statistical terms, we find that low values of the x-axis correspond to high values on the y-axis, and high values on the x-axis correspond to low values on the y-axis.

8.5 No Relationship Displayed in a Scatterplot

Figure 8.3 displays an instance in which the independent and dependent variable are unrelated to one another. In more detail, the graph highlights that the affluence of a country is not related to the population density of that country.

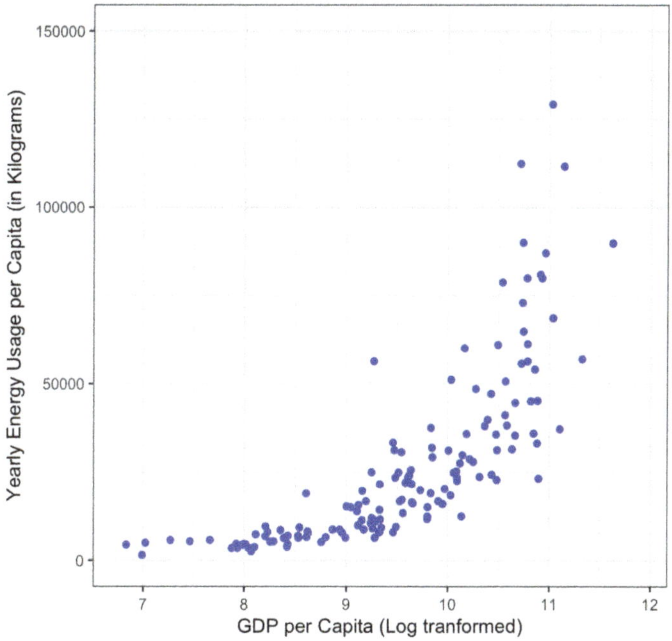

Fig. 8.1 Bivariate relationship between GDP per capita and energy consumption

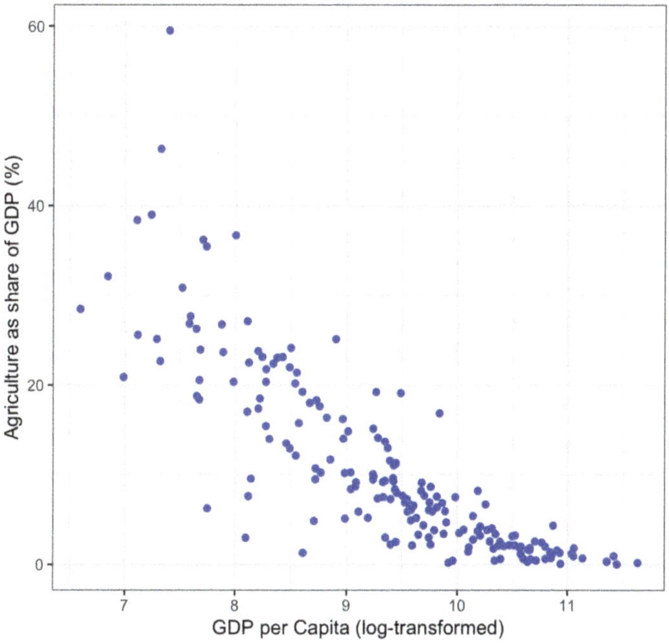

Fig. 8.2 Bivariate relationship between GDP per capita and agriculture as % of GDP

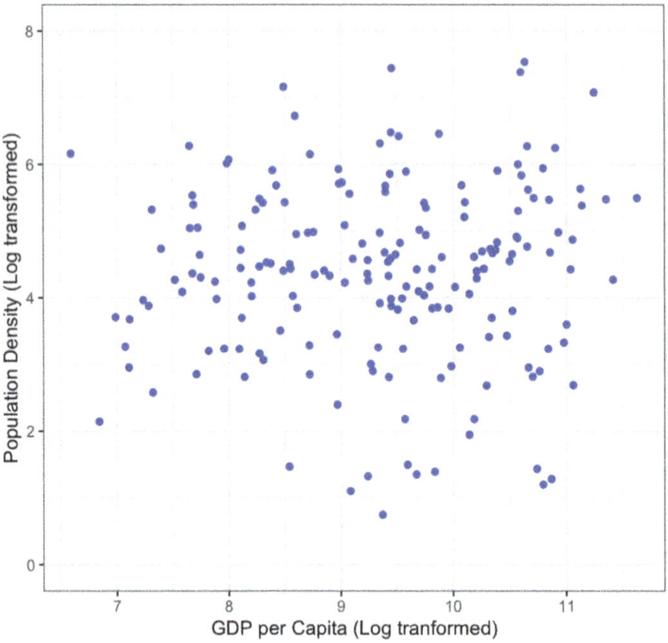

Fig. 8.3 Relationship between a country's GDP per capita and population density

8.6 Drawing a Line in a Scatterplot

The line we draw in a scatterplot is called the ordinary least squares line (OLS line). In theory, we can draw a multitude of lines, but in practice, we want to find the best fitting line to the data. The best fitting line is the line where the summed-up distance of the points from below the line is equal to the summed-up distance of the points from above the line. Figure 8.4 clearly shows a line that does not fit the data properly. The distance of all the points toward the line is much larger for the points below the line in comparison with the points above the line. In contrast, the distance of the points toward the line is the same for the points above the line as for the points below the line in Fig. 8.5. Hence, the line in Fig. 8.5 is the best fitting line.

8.7 Building a Scatterplot in R

Building a scatterplot using R requires the ggplot2 package (Wickham, 2016). This is the go-to package when it comes to data visualization in R. You can find more information and tutorials on the package Website: https://ggplot2.tidyverse.org/index.html.

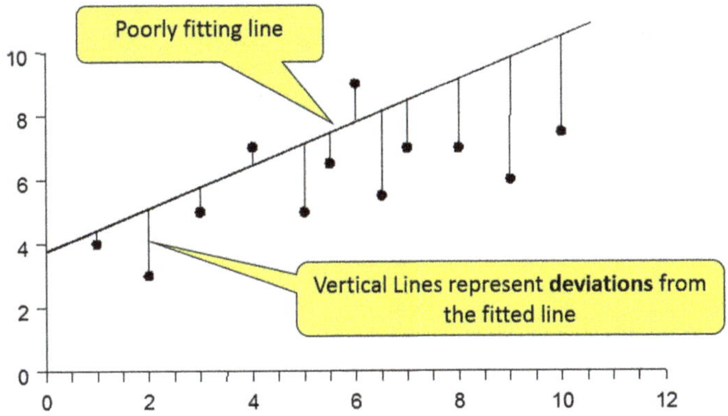

Fig. 8.4 Example of a poorly fitted line

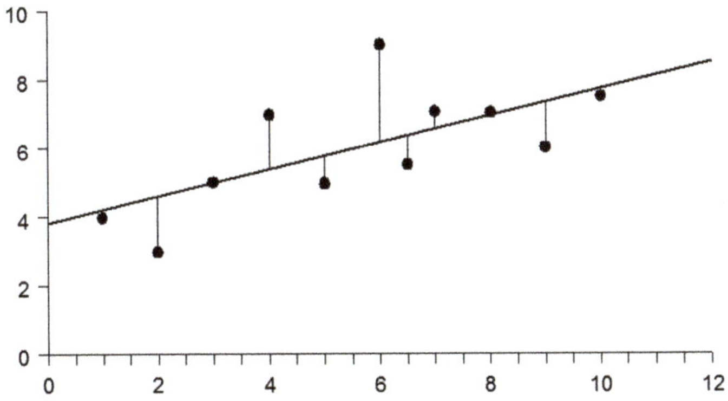

Fig. 8.5 Example of the best fitted line

For our scatterplot, we will use two variables from our party dataset. The dependent variable is money spent partying per week, and the dependent variable is the quality of extracurricular activities. We are hypothesizing that students who enjoy the university sponsored activities will spend less money partying. Rather than going out and party, they will be in sports, social, or political university clubs and partake in extracurricular activities. A scatterplot can help us confirm or disconfirm this hypothesis.

```
## First, we must download the ggplot2 package
install.packages("ggplot2")
library(ggplot2)
```

```
## Now we can build our scatterplot by specifying two arguments. First, the ggplot() argument will help us set up the relationship we wish to plot (our DV and IV). Then, we will specify which type of graph we want to build using the geom_point() argument (this argument builds a scatterplot). We will also add an ordinary least squares line to our scatterplot using the command geom_smooth(method=lm). This specifies that we want a smooth line using a linear model. As you can see from the command below, we are also adding labels for our axes with the argument labs() as well as specifying colors for both the scatterplot (blue) and the OLS line (red). We are also changing the theme of the plot to black and white using the command theme_bw(). This makes the plot easier to read.
ggplot(aes(x = QECA, y = MSP), data = dat) +
  geom_point(colour = "blue") +
  geom_smooth(method = lm, colour = "red") +
  labs(x = "Quality of Extra-Curricular Activities (0–100 scale)",
      y = "Money Spent Partying per week (in $)") +
  theme_bw()

## The scatterplot should look like the one in Fig. 8.6.
```

As expected, Fig. 8.6 displays a negative relationship between the quality of extracurricular activities and the money students spent partying. The graph displays that students who think that the extracurricular activities offered by the university are poor do in fact spend more money per week partying. In contrast, students who like the sports, social, and political clubs at their university are less likely to spend a lot of money partying. Because we can see that the line is relatively steep, we can already detect that the relationship is relatively strong; a bivariate regression analysis (see below) will give us some information about the strength of this relationship.

8.8 Correlation Analysis

A correlation analysis is closely linked to the analysis of scatterplots. In order to do a correlation analysis, the relationship between independent and dependent variable must be linear. In other words, we must be able to use a line to express the relationship between the independent variable x and the dependent variable y. To interpret a scatterplot, two things are important: (1) the direction of the line (i.e., a relationship can only exist if the fitted line is either positive or negative) and (2) the closeness of the points toward the line (i.e., the closer the points are clustered around the line, the stronger the correlation is). In fact, in a correlation analysis, it is solely the second point, the closeness of the points to the line that helps us determine the strength of the relationship. To highlight, if we move from graph 1 to graph 4 in Fig. 8.7, we can see that the points get further apart for each of the

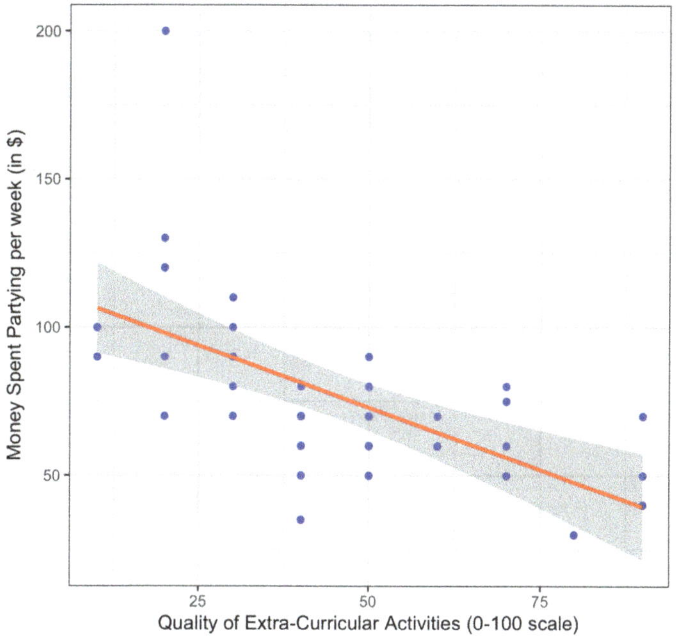

Fig. 8.6 Relationship of quality of extracurricular activities and money spent partying

four graphs, showing that the correlation between the two variables is becoming weaker until it is inexistent in the fourth graph. Similarly, if we move from graph 4 to the last graph, we can see that the points get closer together, showing that the correlation between the two variables becomes stronger.

In statistical terms, the correlation coefficient (also called Pearson correlation coefficient) is denoted by the letter r and ranges from -1 to 1. It expresses both the ***strength*** and ***direction*** of a relationship between two variables. If the dots line up to exactly one line, we have a perfect correlation or a correlation coefficient of 1 (or -1 if negative). In contrast, the more scattered the dots are the more the correlation coefficient will approach 0. In terms of direction, a correlation coefficient with a positive sign depicts a positive correlation, whereas a correlation coefficient with a negative sign depicts a negative correlation.

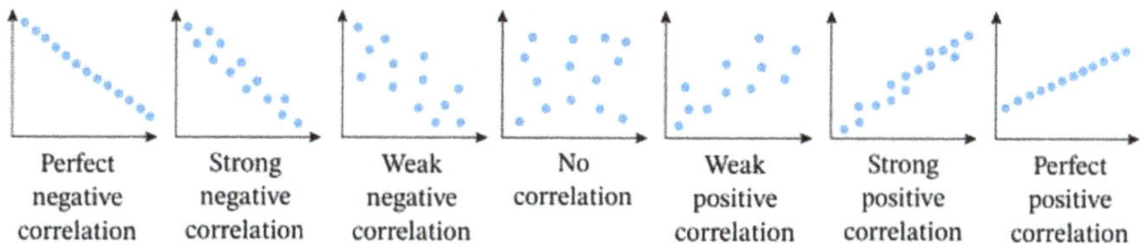

Fig. 8.7 Assessing the strength of relationships in correlation analysis

Table 8.1 Benchmarks for establishing the strength of correlation

Value of coefficient	Strength of correlation
$0.30 < r < 0.45$	Weak correlation
$0.45 < r < 0.60$	Medium/strong correlation
$r < 0.60$	Strong correlation

Note Valid for positive and negative correlations alike

Formula for correlation coefficient:

$$r = \frac{\sum (x_i - \overline{x})(y_i - \overline{y})}{\sqrt{\sum (x_i - \overline{x})^2 \sum (y_i - \overline{y})^2}}$$

Properties of correlation coefficient:

- $-1 < r < 1$
- $r > 0$ means a positive relationship—The relationship is stronger the closer r is to 1
- $r < 0$ means a negative relationship—The relationship is stronger the closer r is to -1
- $r = 0$ means no relationship.

Table 8.1 presents the typical benchmarks for interpreting the strength of a correlation coefficient. These values are valid for both positive and negative coefficients. As we can see, coefficients between 0.30 and 0.45 indicate a weak correlation. Meanwhile, coefficients between 0.45 and 0.60 indicate a medium correlation. Lastly, any coefficient above 0.60 is considered strong.

The correlation coefficient, like the mean and the standard deviation, is sensitive to outliers. For example, Fig. 8.8 displays nearly identical scatterplots, with the sole difference being that the first graph has an outlier, while the second one does not. Removing the outlier increases the correlation coefficient by 0.3 points. Given that the line is drawn so that the sum of the points below the line and the sum of the points above the line equal 0, an outlier pushes the line in one direction (in our case downwards), thus increasing the distance of each point toward the line, which, in turn, decreases the strength of the correlation.

There are two important caveats with correlation analysis. First, since a correlation analysis defines strength at how close the points in a scatterplot are toward a line, it does not provide us with any indication of the strength in impact, in the substantial sense, of a relationship of an independent on a dependent variable. Second, a correlation analysis depicts only whether two variables are related and how closely they follow a positive or a negative direction. It does not give us any indication which variable is the cause, and which is the effect.

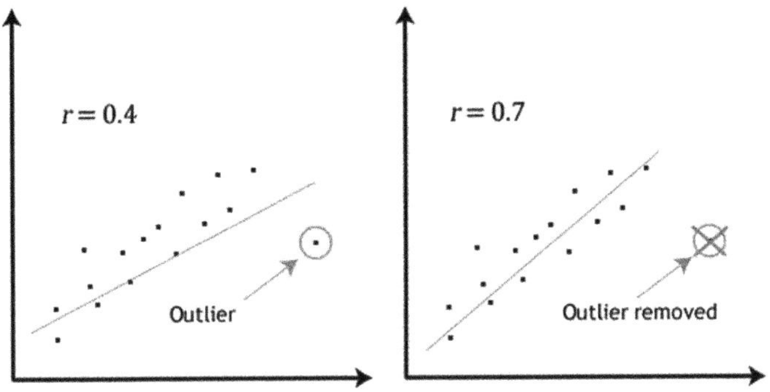

Fig. 8.8 Correlation coefficient with and without an outlier

8.9 Computing a Correlation Analysis in R

In R, we use the command cor() to compute a Pearson correlation coefficient. The following example measures the correlation between the variable quality of extracurricular activities and money spent partying (based on scatterplot in Fig. 8.6).

```
## We can compute the correlation using the cor.test() command. Note that we must specify the type of correlation to Pearson using the 'method = ' argument.
cor.test(dat$QECA, dat$MSP, method = "pearson")

## The output appears below. The result indicates a coefficient of r = –0.60.
Pearson's product-moment correlation

data: dat$QECA and dat$MSP
t = –4.629, df = 38, p-value = 4.2e–05
alternative hypothesis: true correlation is not equal to 0
95 Percent Confidence Interval:
 –0.7682721 –0.3554473
sample estimates:
      cor
–0.6004695
```

8.9.1 Interpreting and Reporting the Results of a Correlation Using R

The correlation output has several important components. First, we have the t-value and *p*-value. In this case, our *p*-value is $p = 0.000042$, which indicates that our correlation is statistically significant and that the relationship between the quality of extracurricular activities and money spent partying is significant. The correlation coefficient tells us the strength and direction of this relationship. Our coefficient ($r = -0.60$) implies a strong negative relationship between quality of extracurricular activities and money spent partying. As we had hypothesized previously, those who indicate that they have better quality extracurricular activities tend to spend less money on partying. Conversely, those who report not having good quality extracurricular activities indicate spending more money on partying. The correlation output also provides us with a 95% confidence interval for the correlation coefficient. If we were to reproduce this correlation in 100 samples, we would obtain a correlation coefficient between −0.36 and −0.77 at least 95 times.

8.10 Bivariate Regression Analysis

In correlation analyzes, we look at the direction of the line (positive or negative) and at how closely the points of the scatterplot follow that line. This allows us to detect the degree to which two variables covary, but it does not allow us to determine how strongly an independent variable influences a dependent variable. In regression analysis, we are interested in the magnitude of the influence of an independent variable on a dependent variable, as measured by the steepness of the slope. To determine the influence of an independent variable on a dependent variable, two things are important: (1) the steeper the slope, the more strongly the independent variable impacts the dependent variable and (2) the closer the points are to the line, the more certain we can be that this relationship actually exists.

8.11 Gauging the Steepness of a Regression Line

To explain the notion that a steeper slope indicates a stronger relationship, let us compare the two graphs in Fig. 8.9. Both graphs depict a perfect relationship, meaning that all the points are on a straight line. The correlation for both would be 1. However, we can see that the first line is much steeper than the second line. In other words, the y value grows stronger with higher x values. In contrast, the second line only moves slightly upwards. Consequently, the regression coefficient is higher in the first compared to the second graph.

To determine the relationship between x and y, we use a point slope:

$$y = a + bx + \epsilon$$

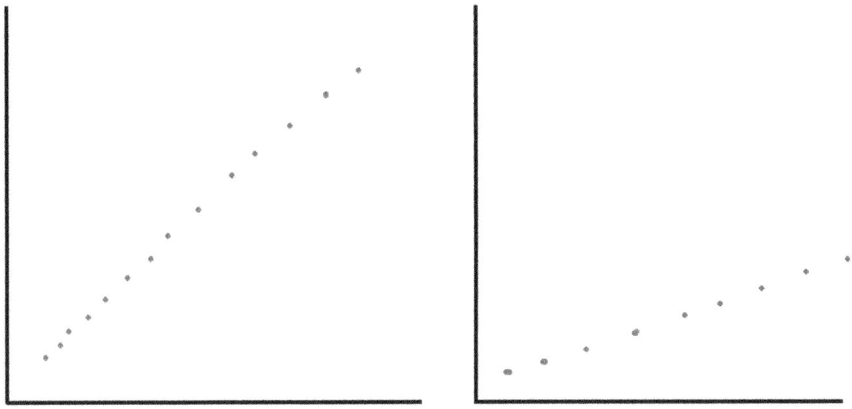

Fig. 8.9 Two regression lines featuring a strong and weak relationship, respectively

where y is the dependent variable
x is the independent variable
b is the slope of the line
a is the intercept (y when x=0)
ϵ is the residual term (see Sect. 8.6.2)
The formula for the regression line:

$$b = r\frac{S_y}{S_x}$$

$$a = \overline{y} - b\overline{x}$$

When we draw the regression line, we could in theory draw and an infinite number of lines. The line that explains the data best is the line that has the smallest sum of squared errors. In statistical terms, this line is called the least square line (OLS line).

Figure 8.10 displays the least square line between the independent variable of self-reported rate of seat belt usage and the dependent variable number of road fatalities per million of traffic miles. The equation for this relationship is

$$y = 0.026 - 0.012x$$

where y is the number of car crash fatalities per million of traffic miles
x is the rate of seat belt usage (self-reported)
This means that there are 0.026 fatalities per million of traffic miles when no one wears their seatbelts. The equation further predicts that for every 1 unit increases on the x-axis (that is, every 1 unit increases in the rate of seat belt usage), y decreases by 0.012. In this example, this means that if everyone reported wearing their seatbelt, the number of road accident fatalities would decrease by 0.012.

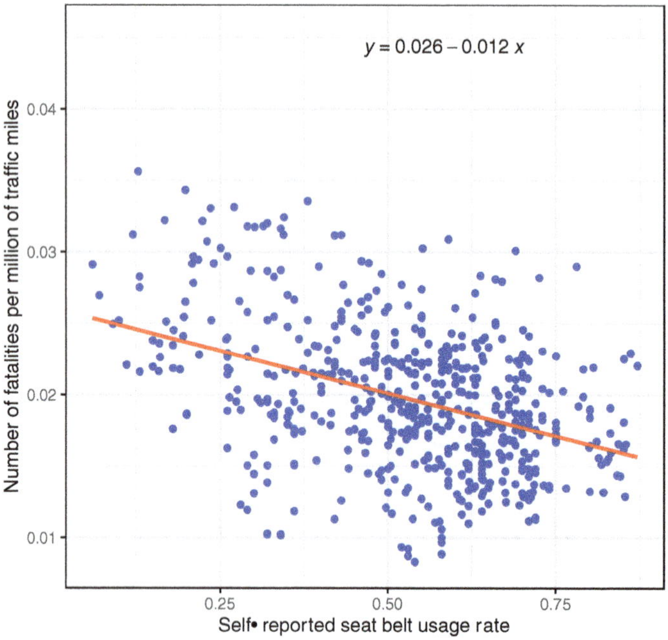

Fig. 8.10 Regression line and equation of seat belt usage and road fatalities

Figure 8.10 illustrates this relationship with the red line (representing the OLS regression line). In this graph, the shaded area either side of the line represents the standard error of the regression.

8.12 Gauging the Error Term

The metric we use to determine the magnitude of a relationship between an independent and a dependent variable is the steepness of the slope. As a rule, we can say that the steeper the slope, the more certain we can be that a relationship exists. However, the steepness of the slope is not the only criterion we use to determine whether or not an independent variable relates to a dependent variable. Rather, we must also look how close the data points are toward the line. The closer the points are toward the line, the less error there is in the data. Figure 8.11 displays two identical lines measuring a relationship between x and y. What we can see is that the relationship is equally strong for the two lines (same regression line). However, the data fits the second line (Example 2) much better than the first line (Example 1). There is more 'noise' in the data in the first example, thus rendering the estimation of each of the points or observations less exact in the first example when compared to the second one.

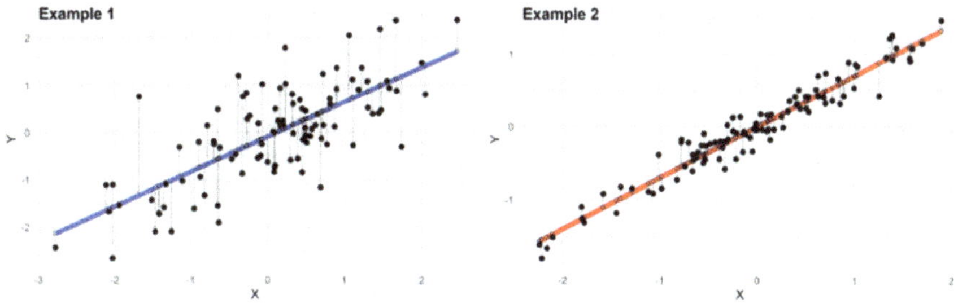

Fig. 8.11 Example of OLS regressions with large and small standard error, respectively

Figure 8.12 graphically explains what we mean by error term or residual. The error term or residual in a regression analysis measures the distance from each data point to the regression line. The farther any single observation or data point is away from the line, the less this observation fits the general relationship. The larger the average distance is from the line, the less well does the average data point fits the linear prediction. In other words, the greater the distance between the average data point and the line, the less we can be assured that the relationship portrayed by the regression line actually exists.

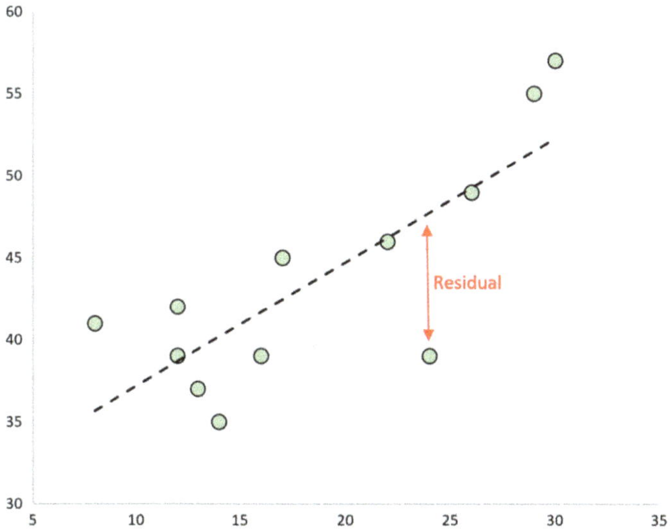

Fig. 8.12 Error term (residual) in a regression analysis

8.13 Computing a Bivariate Regression Analysis in R

To conduct a bivariate regression analysis, we take the same variables we used for the scatterplots, that is, our dependent variable is money spent partying, and the independent variable is the quality of extracurricular activities.

```
## We can compute the correlation using the lm() command (lm for linear model). We specify our regression equation (y ~ x) and where the data comes from (dat). Notice that unlike the previous analyzes, we must assign our regression model to a given named vector (in this case OLSmodel) which we can then summarize using the function summary().
OLSmodel < - lm(QECA ~ MSP, data = dat)
summary(OLSmodel)

## The output appears below.

Call:
lm(formula = MSP ~ QECA, data = dat)

Residuals:
    Min      1Q  Median      3Q     Max
-46.322 -12.936  -3.742  10.371 101.904

Coefficients:
            Estimate Std. Error t value Pr(>|t|)
(Intercept) 114.8693     9.1446  12.561 4.22e-15 ***
QECA         -0.8387     0.1812  -4.629 4.20e-05 ***
--
Signif. codes:  0 '***' 0.001 '**' 0.01 '*' 0.05 '.' 0.1 ' ' 1

Residual standard error: 24.43 on 38 degrees of freedom
Multiple R-squared: 0.3606, Adjusted R-squared: 0.3437
F-statistic: 21.43 on 1 and 38 DF, p-value: 4.2e-05
```

8.14 Interpreting the Regression Output

8.14.1 Regression Coefficient and Intercept Estimates

In the regression model output above, we can see that our regression coefficient (estimate) is −0.84 and is statistically significant at $p = 0.000042$. This coefficient implies that for each one unit increases in the quality of extracurricular activities (on the 0–100 scale), students spend on average 0.84$ less on partying. Our intercept estimate indicates that a student who reports a score of 0 on the quality of extracurricular activities scale spends, on average, 114.87$ on partying every week. From the parameters in the regression output, we can build our regression line equation:

$$\text{Money spent partying} = 114.87\$ + -0.84 * \text{Quality of extracurricular activities}$$

We can use this equation to determine the expected amount of money a student would spend on partying per week based on that student's score on the quality of extracurricular activities scale (QECA scale). For example, a student who indicates a score of 75 on the QECA scale would spend, on average, 51.87$ per week on partying.

8.15 Standard Error and t-Value

The regression model output in R also provides us with two important parameters: the standard error of the coefficient estimate and the t-value. These coefficients have important implications.

The **standard error** gives us an indication of how much variation there is around the predicted coefficient. As a rule, we can say that the smaller the standard error in relation to the regression weight is, the more certainty we can have in the interpretation of our relationship. In our case, the standard error behind the relationship between extracurricular activities and money spent partying is 0.181.

The **t-value** compares the unstandardized regression against a predicted value of zero (i.e., no contribution to regression equation). It does so by dividing the unstandardized regression coefficient (i.e., our measure of effect size) by the standard error (i.e., our measure of variability in the data). As a rule, we can say that the higher the t-value, the higher the chance that we have a statistically significant relationship (see significance value). In our example, the t-value is -4.629, which is statistically significant.

8.16 Model Fit

The regression model output also indicates how well the model fits the data. It consists of four parameters: (1) the residual standard error, (2) the R-squared value, (3) the adjusted R-squared value, and (4) the F-statistic.

The **residual standard error** is the standard deviation of the error term and the square root of the mean square residual (or error). Normally, we do not interpret this estimator when we conduct regression models. In our sample, the standard error of the estimate is 24.43.

R-squared is the most important parameter in the model summary output. It is a measure of model fit (i.e., it is the squared correlation between the model's predicted values and the real values). It explains how much of the variance in the dependent variable the independent variable(s) in the model explain. In theory, the R-squared values can range from 0 to 1. An R-squared value of 0 means that the independent variable(s) do not explain any of the variance of the dependent variable, and a value of 1 signifies that the independent variable(s) explain all the variance in the dependent variable. In our example, the R-squared value of 0.361 implies that the independent variable—the quality of extracurricular activities—explains 36.1% of the variance in the dependent variable—the money students spent partying per week.

The **adjusted R-squared** is a second statistic of model fit. It helps us compare different models. In real research, it might be helpful to compare models with a different number of independent variables to determine which of the alternative models is superior in a statistical sense. To highlight, the R-squared will always increase or remain constant if I add variables. Yet, a new variable might not add anything substantial to the model. Rather, some of the increase in R-squared could be simply due to coincidental variation in a specific sample. Therefore, the adjusted R-squared will be smaller than the R-squared since it controls for some of the idiosyncratic variance in the original estimate. Therefore, the adjusted R-squared is a measure of model fit adjusted for the number of independent variables in the model. It helps us compare different models; the best fitting model is always the model with the highest adjusted R-squared (not the model with the highest R-squared).[1] In our sample, the adjusted R-squared is 0.344. We do not interpret this estimator in bivariate regression analysis and only use it if we compare various models.

The **F-statistic** in a regression model works like an F-test or ANOVA. It is an indication of the significance of the model (does the regression equation fit the observed data adequately?). If the F-statistic is significant, the regression equation has predictive power, which means that we have at least one statistically significant variable in the model. In contrast, if the F-test is not significant, then none of the variables in the model is statistically significant, and the model has no predictive power. In our sample, the F-statistic is 21.43, and the corresponding significance level is $p = 0.000042$. Hence, we can already conclude that our independent variable (the quality of extracurricular activities) influences our dependent variable (the money students spend per week partying).

[1] Please note that we can only compare adjusted R-squared values of different models if these models have the same number of observations.

8.17 Reporting Regression Results with a Model Table

When we report the results of a bivariate regression model, we report the coefficient, standard error, and significance level, as well as the R-squared and the number of observations in the model. We would also accompany the interpretation of the results with a regression table. In R, we can build this table using several different packages. In this example, we use the modelsummary package to build an elegant table presenting the results of our regression model (Arel-Bundock, 2022). After building the table, we can present the results as we would in a peer-reviewed academic publication or a research paper.

```
## Let's build our regression model table using modelsummary()
modelsummary(OLSmodel, stars = TRUE)

## The generated output (Table 23) is presented below along with our results
(as we would in a publication).
```

8.18 Presenting the Results in a Research Article

Table 8.2 reports the results of a bivariate OLS regression model measuring the influence of the quality of extracurricular activities on students' weekly spending patterns for partying based on a survey conducted with 40 undergraduate students at a Canadian university. The model portrays a negative and statistically significant relationship between the two variables. In substantive terms, the model predicts that for every point students' evaluation of the extracurricular activities at their university increases, they spent 84 cents less per week partying. This influence is substantial. For example, somebody who thinks that the extracurricular activities at her university are poor (and rates them at 20) is predicted to spend approximately 98 dollars for partying activities. In contrast, somebody, who likes the extracurricular activities (and rates them at 80) is only expected to spend 48 dollars for her weekly partying. The adjusted R-squared of the model further highlights that the independent variable, the quality of extracurricular activities, explains 34 percent of the variance in the dependent variable money spent partying per week.

Table 8.2 OLS regression results for quality of extracurricular and money spent partying

	Money spent partying
(Intercept)	114.869***
	(9.145)
QECA	−0.839***
	(0.181)
N	40
R^2	0.361
Adjusted R^2	0.344
F	21.427

Note Unstandardized regression coefficient with standard error in parentheses. Table created using modelsummary (Arel-Bundock, 2022)
* $p < 0.05$, ** $p < 0.01$, *** $p < 0.001$

References

Arel-Bundock, V. (2022). modelsummary: Data and model summaries in R. *Journal of Statistical Software, 103*(1), 1–23. https://doi.org/10.18637/jss.v103.i01

Wickham, H. (2016). *ggplot2: Elegant graphics for data analysis.* Springer-Verlag.

Further Readings

Gravetter, F. J., & Forzano, L. A. B. (2018). *Research methods for the behavioral sciences.* Cengage Learning (Chap. 12).

Nice introduction into correlational research; covers the data and methods for correlational analysis, applications of the correlational strategy, and strength and weakness of the correlational research strategy.

Montgomery, D. C., Peck, E. A., & Vining, G. G. (2012). *Introduction to linear regression analysis* (Vol. 821). Wiley & Sons (Chaps. 1 and 2).

Chapters 1 and 2 provide a hands-on and practical introduction into linear regression modelling, first in the bi-variate realm and then in the multivariate realm.

Ott, R. L., & Longnecker, M. T. (2015). *An introduction to statistical methods and data analysis.* Nelson Education (Chap. 11).

Provides a good introduction into correlation and regression analysis clearly highlighting the differences between the two techniques.

Roberts, L. W., Wilkinson, L., Peter, T., & Edgerton, J. (2015). *Understanding social statistics.* Oxford University Press (Chaps. 10–14).

These chapters offer students the basic tools to examine the form and strength of bivariate relationships.

Multivariate Regression Analysis

9

Bivariate regression analysis is very rarely used in real applied research, because an outcome is hardly ever just dependent on one predictor. Rather, multiple factors normally explain the dependent variable (see Fig. 9.1). To highlight, if we want to explain a student's grade in an exam, several factors might come into play. A student's grade might depend on how much the respective student studied for the exam, it might depend on her health, and even on her general mood. Multiple regression modelling allows us to comparatively gauge the influence of all of these factors on the dependent variable.

Multiple regression analysis is an extension of bivariate regression analysis. It allows us to test the influence of multiple independent variables (predictors) on a dependent variable. Just like in the case of two variables, the goal of this method is to create an equation or a "model" that explains the relationship between these variables.

Let us assume that we want to explain the dependent variable "y", and we have several independent variables (x_1, x_2, x_3, ..., x_p). Then, the multiple regression equation we need to calculate is:

$$y' = a + \beta_1 x_1 + \beta_2 x_2 + \ldots + \beta_p x_p + \epsilon$$

where

y' is the predicted value of the dependent variable
x_p are the independent variables
β_p are the slopes for each independent variable
a is the y-intercept (value of y' when all values of x are equal to 0)
ϵ is the residual term

© The Author(s), under exclusive license to Springer Nature Switzerland AG 2023

D. Stockemer and J.-N. Bordeleau, *Quantitative Methods for the Social Sciences*, Springer Texts in Political Science and International Relations,
https://doi.org/10.1007/978-3-031-34583-8_9

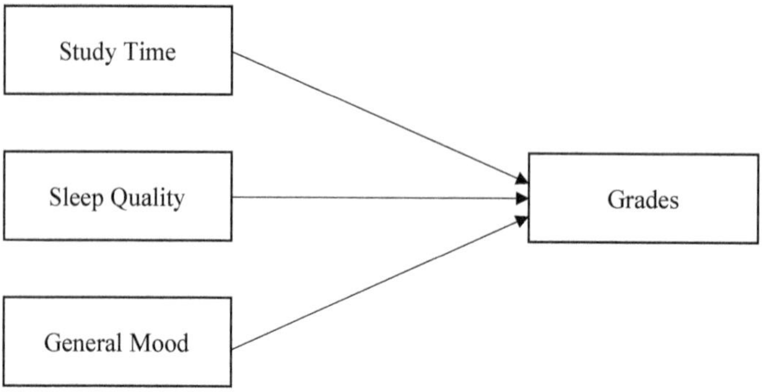

Fig. 9.1 Predictors of a student's grades

Example Suppose we want to study the predictors of a student's grade in a math exam (see Fig. 9.1). We ask a random sample of students at a German high school about their grade in their last math exam, the time they spent studying for this exam, their general mood and whether they were in good perceived health when taking the exam. Let us further assume that we run a multiple regression model and receive the following equation (for now we ignore the question of whether the variables are statistically significant or not):

$$y' = 10.5 + 3.1x_1 + 1.5x_2 + 0.5x_3 + \epsilon$$

We would interpret the equation as follows:

- **10.5** (on a scale from 0 to 100) is the hypothetical grade a student is expected to get, if she does not study at all, her general health is at its worst (she would rank her health by the value 0) and her general mood is also at the lowest value (she would also rank this at 0).
- **3.1** is the slope coefficient for the variable study time. This implies that for every hour a student studies, her grade is expected to increase by 3.1 points.
- **1.5**. is the slope coefficient for somebody's general health, indicating that per each point somebody's perceived health increases, her math grade is predicted to improve by 1.5 points.
- **0.5** is the slope coefficient for somebody's general mood. In other words, for each point somebody's general mood increases, her test performance is expected to increase by 0.5 points.

As we can see from the example, the multivariate regression model is an extension of the bivariate model. It has the same parameters, and the interpretation is analogous. To do a multiple regression analysis in R, we follow the same steps as we would follow for a bivariate regression. We simply add more independent variables in the equation.

9.1 The Forms of Independent Variables to Include into Multivariate Regression Models

In this introduction to survey research and quantitative methods, we only cover continuous dependent variables (non-continuous dependent variables will be the subject of more advanced statistical courses). With that being said, we must still determine in which type of functional form we would include our independent variables. Provided that the relationship between a continuous independent and a continuous dependent variable is linear (the relationship follows a line), we will include the independent variables in its linear form. If a scatterplot were to highlight that the relationship between independent and dependent variable is not linear (e.g., it follows a curve), we would need to transform that variable. However, this is also material for a more advanced textbook and course.

For the purpose of this introductory textbook, the only variables we must be careful with are categorical variables. If we have a categorical nominal variable (i.e., different religious affiliations), we create $N - 1$ dummy variables with one of the categories serving as a reference category. If we have ordinal variables, we could also test whether the relationship is in fact ordered. For example, we could test via a multiple comparison test whether the relationship between times partying per week and money spent partying per week is in principle ordinal. In our example, including the variable times partying per week in its linear ordinal form assumes that the relationship between partying less than one time per week and one time per week is the same as that between four and five times per week. However, in the ANOVA, we find that this is not true. Rather, we find that students that party three times or less spend approximately the same amount of money when they party; only students that party four or more times spend significantly more. Because of this dichotomy in the relationship, it would make sense to create a dummy variable between the two categories (in the dataset we label this variable TP3). The code to create this variable is presented below:

```
## Let's create the dichotomous variable TP3.
dat$TP3 <- recode(
  dat$TP, '0' = 0, '1' = 0, '2' = 0, '3' = 0, '4' = 1, '5' = 1)
```

9.2 Interpreting a Multivariate Regression Model

When we want to interpret a multivariate regression model, we can follow the same logic as for a bivariate regression model. The four guiding steps can help:

1. Look at what variables are significant.
2. Interpret the substantive value of significant variables.

3. Compare the relative strength of the significant variables.
4. Interpret the model fit.

9.3 Computing a Multiple Regression Model in R

In our sample survey, we have included seven possible predictors and we want to determine the relative and absolute influence of these seven variables on the dependent variable (money spent partying per week). Because we know from the ANOVA analysis that the relationship between the ordinal variable Times Partying and Money Spent Partying is not linear, we create a binary variable, coded 0 for partying three times or less per week and 1 for partying 4 times or more (see code in Sect. 9.2). We add this recoded independent variable together with the remaining six independent variables into the model (see Sect. 9.7) and label it TP3. The dependent variable is money spent partying per week.

```
## Let's compute a multiple regression model with our seven predictors and
the dependent variable money spent partying per week.
  multipleregression <- lm(MSP ~ ST + Gender + Year + TP3 + FWA +
QECA + ATSP,
                          data = dat)
  summary(multipleregression)

  ## The output appears below.

  Call:
  lm(formula = MSP ~ ST + Gender + Year + TP3 + FWA + QECA +
ATSP,
     data = dat)

  Residuals:
     Min      1Q  Median     3Q    Max
  −42.171 −10.816  −0.888  8.786 76.628

  Coefficients:
              Estimate Std. Error t value Pr(>|t|)
  (Intercept) 75.46679   34.62063   2.180  0.03674 *
  ST           1.37069    2.18487   0.627  0.53488
  Gender      −3.31611    8.29012  −0.400  0.69181
  Year         1.77135    4.15203   0.427  0.67251
  TP3         26.64110   10.37290   2.568  0.01509 *
  FWA         −0.05834    0.28365  −0.206  0.83834
```

```
QECA        -0.58836   0.17983  -3.272  0.00256 **
ATSP         0.17889   0.11943   1.498  0.14396
---
Signif. codes: 0 '***' 0.001 '**' 0.01 '*' 0.05 '.' 0.1 ' ' 1

Residual standard error: 21.75 on 32 degrees of freedom
Multiple R-squared: 0.5731, Adjusted R-squared: 0.4797
F-statistic: 6.137 on 7 and 32 DF, p-value: 0.0001347
```

9.4 Interpreting a Multiple Regression Model

Following the four steps outlined under Sect. 9.3, we can proceed as follows (see Table 9.1):

Table 9.1 OLS multiple regression results

	Money spent partying
(Intercept)	75.467*
	(34.621)
Study time	1.371
	(2.185)
Gender	−3.316
	(8.290)
Year	1.771
	(4.152)
Times partying (dichotomous)	26.641*
	(10.373)
Fun without alcohol	−0.058
	(0.284)
Quality of extracurricular activities	−0.588**
	(0.180)
Average tuition student pays	0.179
	(0.119)
N	40
R^2	0.573
Adjusted R^2	0.480
F	6.137

Note Unstandardized regression coefficients with standard error in parentheses. Table created using model summary (Arel-Bundock, 2022)

* $p < 0.05$, ** $p < 0.01$, *** $p < 0.001$

1. If we look at the significance level, we find that two variables are statistically significant (i.e., the quality of extracurricular activities and our dichotomous times partying variable). For all other variables the significance level is higher than the benchmark of $p < 0.05$. Thus, we would conclude that these indicators do not influence the amount of money students spend partying per week.
2. The first significant variable (the quality of extracurricular activities) has the expected negative sign indicating that the students who enjoy the extracurricular activities at their institution spend less money partying each week. This observation also confirms our initial hypothesis. Holding everything else constant, the model predicts that per every point increase on the quality of extracurricular activities scale, she spends 58 cents less per week partying.
3. The second significant variable (TP3) also has the expected positive sign. The regression coefficient 26.64 indicates that people that party 4 or more times are expected to spend nearly 27 dollars more on their partying habits per week than students that party three times or less.
4. The model fits the data quite well; the seven independent variables explain 57% of the variance in the dependent variable, the amount of money students spend partying per week (R-squared = 0.57).

9.5 Reporting the Results of a Multiple Regression Analysis

In the multiple regression analysis (see Table 9.1), we evaluated the influence of seven independent variables on the weekly amount of money students spend partying. We find that two of the seven predictors are statistically significant and show the expected effect; that is the more students think that the extracurriculars at their university are good, the less money they spend partying per week. The same applies to students that party few times, they too spend less money going out. In substantive terms, the model predicts that for every point increase in students' ranking of the extracurricular activities at their school, they will spend 59 cents less partying per week. The coefficient for the times partying per week variable indicates that students who party three or more times are predicted to spend 26 dollars more on their partying habits than students that party less. Relying on an alpha benchmark of 0.05, none of the other variables are statistically significant. In terms of model fit, the data fits the model fairly well: the seven independent variables explain 57 percent of the variance in the dependent variable.

9.6 Finding the Best Model

In academic research, the inclusion of variables into a regression model should be theoretically driven; that is theory should tell us which independent variables to include in a model to explain and predict a dependent variable. However, we might also be interested in finding the best model based on diverging theoretical views. There are two ways to proceed, and there is some disagreement among statisticians: One way is to only include statistically significant variables into the model. Another way is to use the adjusted *R* squared as a benchmark. To recall, the adjusted *R*-squared is a measure of model fit that allows us to compare different models. For every additional predictor I include in the model, the adjusted *R* squared increases only if the new term improves the model beyond pure chance (note that a poor predictor can decrease the adjusted R-squared, but it can never decrease the *R*-squared). Using the adjusted *R*-squared as a benchmark to find the best model, we should proceed as follows: (1) start with the complete model, which includes all the predictors, (2) remove the non-statistically significant predictor with the lowest standardized coefficient, (3) continue this procedure until the adjusted R-squared no longer increases.

Table 9.2 highlights this procedure. We start with the full model. The full model has an adjusted *R* squared of 0.480. We take out the variable with the lower unstandardized beta coefficient (fun without alcohol). After taking out this variable, we see that the adjusted *R* squared increases to 0.495 (see Model 2). This indicates that the variable fun without alcohol does not add anything substantial to the model and should be removed. In the next step, we remove the variable year of study. Removing this variable leads to another increase in the adjusted *R*-squared (i.e., the new adjusted *R*-squared is 0.508), indicating again that this variable does not add anything substantively to the model and should be removed (see Model 3). Next, we remove the variable gender and see another increase in the adjusted *R*-squared to 0.519. If we now remove the variable with the lowest regression coefficient (study time per week), we find that the adjusted *R*-squared decreases to 0.513 (see Model 5), which is lower than the adjusted *R*-squared from Model 4. Based on this approach, we can conclude that Model 4 has the best model fit.

9.7 Assumptions of the Ordinary Least Squares Regression Model (OLS)

The ordinary least-squares linear regression model (OLS) is the simplest type of regression model. OLS models only works with a continuous dependent variable. It has 10 underlying assumptions:

1. **Linearity in the parameters**: linearity in the parameters implies that the relationship between a continuous independent variable and a dependent variable must roughly follow a line. Relationships that do not follow a line (e.g., they might follow a quadratic function or a logarithmic function) must be included into the model using the correct functional forms (more advanced textbooks in regression analysis will capture these cases).
2. **X is fixed**: this rule implies that one observation can only have one X and one Y value.

Table 9.2 Finding the model with the best fit

	Model 1	Model 2	Model 3	Model 4	Model 5
(Intercept)	75.467*	70.225**	76.360***	**77.527*** **	92.663***
	(34.621)	(23.090)	(16.275)	**(15.810)**	(9.926)
Study time	1.371	1.726	1.546	**1.367**	
	(2.185)	(1.321)	(1.217)	**(1.116)**	
Gender	−3.316	−3.672	−3.047		
	(8.290)	(7.989)	(7.719)		
Year	1.771	1.377			
	(4.152)	(3.628)			
Times partying	26.641*	27.584**	28.973**	**28.772** **	28.765**
	(10.373)	(9.169)	(8.300)	**(8.184)**	(8.240)
Fun without alcohol	−0.058				
	(0.284)				
Quality of extracurriculars	−0.588**	−0.580**	−0.581**	**−0.588** **	−0.618***
	(0.180)	(0.172)	(0.170)	**(0.167)**	(0.167)
Average tuition paid	0.179	0.184	0.170	**0.152**	0.127
	(0.119)	(0.115)	(0.108)	**(0.097)**	(0.095)
N	40	40	40	**40**	40
R^2	0.573	0.573	0.571	**0.569**	0.550
Adjusted R^2	0.480	0.495	0.508	**0.519**	0.513
F	6.137	7.367	9.039	**11.538**	14.680

Note Unstandardized regression coefficients with standard error in parentheses. Table created using model summary (Arel-Bundock, 2022)
+ $p < 0.1$, * $p < 0.05$, ** $p < 0.01$, *** $p < 0.001$

3. **Mean of disturbance is zero**: this follows the rule of how we draw the ordinary least squares line. We draw the best fitting line, which implies that the summed-up distance of the points below the line is the same as the summed-up distance above the line.
4. **Homoscedasticity**: the homoscedasticity assumption implies that the variance around the regression line is similar for all the predictor variables around the regression line (X) (see Fig. 9.2). To highlight, in the first graph the points are distributed rather equally around a hypothetical line. In the second graph, the points are closer to the hypothetical line at the bottom of the graph in comparison to the top of the graph. In our example, the first graph would be an example of homoscedasticity and the second graph an example of data suffering from heteroscedasticity. At this stage in your learning, it is important that you have heard about heteroscedasticity, but details of the problem will be covered in more advanced textbooks and classes.

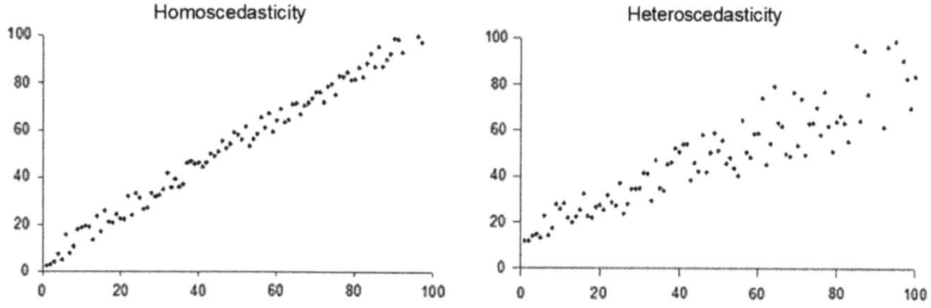

Fig. 9.2 Homoscedasticity and heteroscedasticity

5. **No autocorrelation**: there are basically two forms of autocorrelation: (1) contemporaneous correlation, where the dependent variable from one observation affects the dependent variable of another observation in the same dataset (e.g., Mexican growth rates might not be independent because growth rates in the US might affect growth rates in Mexico). (2) Autocorrelation in pooled time series datasets. That is, past values of the dependent variable influence future values of the dependent (e.g., the US's growth rate in 2017 might affect the US's growth rate in 2018). This second type of autocorrelation is not really pertinent for cross-sectional analysis but becomes relevant for panel analysis.
6. **No endogeneity**: endogeneity is one of the fundamental problems in regression analysis. Regression analysis assumes that the independent variable impacts the dependent variable but not vice versa. In many real-world political science scenarios, this assumption is problematic. For example, there is debate in the literature whether high women's representation in elected office influences/decreases corruption or whether low levels of corruption foster the election of women (see Esarey & Schwindt-Bayer, 2017). There are statistical remedies such as instrumental regression techniques, which can model a feedback loop, that is, more advanced techniques can measure whether two variables influence themselves mutually. These techniques will also be covered in more advanced books and classes.
7. **No omitted variables**: we have an omitted variable problem if we do not include a variable in our regression model that theory tells us that we should include. Omitting a relevant or important variable from a model can have four negative consequences: (1) if the omitted variable is correlated with the included variables, then the parameters estimated in the model are biased, meaning that their expected values do not match their true values. (2) the error variance of the estimated parameters is biased. (3) The confidence intervals of included variables and more general the hypothesis-testing procedures are unreliable, and (4) the *R* squared of the estimated model is unreliable.
8. **More cases than parameters (N > k)**: technically, a regression analysis only runs if we have more cases than parameters. In more general terms, the regression estimates become more reliable the more cases we have.

9. **No constant "variables"**: for an independent variable to explain variation in a dependent variable, there must be variation in the independent variable. If there is no variation, then there is no reason to include the independent variable in a regression model. The same applies to the dependent variable. If the dependent variable is constant or near constant, and does not vary with independent variables, then there is no reason to conduct any analysis in the first place.
10. **No perfect collinearity among regressors**: this rule means that the independent variables included in a regression should represent different concepts. To highlight, the more two variables are correlated, the more they will take explanatory power from each other (if they are perfectly collinear, a regression program such as R cannot distinguish these variables from one another). This becomes problematic because relevant variables might become non-significant in a regression model, if they are too highly correlated with other relevant variables. More advanced books and classes will also tackle the problem of perfect collinearity and multicollinearity. For the purposes of an introductory course, it is enough if you have heard about multicollinearity.

References

Arel-Bundock, V. (2022). modelsummary: Data and model summaries in R. *Journal of Statistical Software, 103*(1), 1–23. https://doi.org/10.18637/jss.v103.i01

Esarey, J., & Schwindt-Bayer, L. A. (2017). Women's representation, accountability and corruption in democracies. *British Journal of Political Science,* 1–32.

Further Readings

Since basically all books listed under bivariate correlation and regression analysis also cover multiple regression analysis, the books we present here go beyond the scope of this textbook. These books could be interesting further reads, in particular to students who want to learn more about what is covered in this textbook and on more advanced topics in regression analyses.

Heeringa, S. G., West, B. T., & Berglund, P. A. (2017). Applied survey data analysis. Chapman and Hall/CRC.

An overview of different approaches to analyze complex sample survey data. In addition to multiple linear regression analysis the topics covered include different types of maximum likelihood estimations such as logit, probit, and ordinal regression analysis, as well as survival or event history analysis.

Lewis-Beck, C., & Lewis-Beck, M. (2015). *Applied regression: An introduction* (Vol. 22). Sage Publications.

A comprehensive introduction into different types of regression techniques.

Pesaran, M. H. (2015). *Time series and panel data econometrics.* Oxford University Press.

Comprehensive introduction into different forms of time series models and panel data estimations.

Wooldridge, J. M. (2015). *Introductory econometrics: A modern approach.* Nelson Education.

Comprehensive book about various regression techniques; it is, however, mathematically relatively advanced.

Appendix 1: The Data of the Sample Questionnaire

	MSP	ST	Gender	Year	TP	FWA	QECA	ATSP
Student 1	50	7	0	2	2	60	90	30
Student 2	35	8	1	3	1	70	40	50
Student 3	120	12	1	3	4	30	20	60
Student 4	80	3	0	4	4	50	50	100
Student 5	100	11	0	1	1	30	10	0
Student 6	120	14	1	5	4	20	20	0
Student 7	90	11	0	4	2	50	50	0
Student 8	80	10	1	4	3	40	50	10
Student 9	70	9	0	3	3	30	50	60
Student 10	80	8	1	2	3	40	40	100
Student 11	60	12	1	4	2	60	40	0
Student 12	50	14	1	2	1	30	70	10
Student 13	100	13	0	3	4	0	30	0
Student 14	90	15	0	3	0	0	20	0
Student 15	60	7	0	3	3	60	50	0
Student 16	40	6	1	4	0	70	90	0
Student 17	60	5	1	4	2	80	50	60
Student 18	90	8	0	5	2	90	30	70
Student 19	130	12	1	4	5	10	20	50
Student 20	70	11	0	1	1	20	60	30
Student 21	80	13	1	3	4	10	70	50
Student 22	50	6	0	4	4	60	40	10
Student 23	110	5	1	4	4	70	30	100
Student 24	60	8	1	2	3	50	60	100
Student 25	70	10	1	4	2	60	40	80

(continued)

© The Editor(s) (if applicable) and The Author(s), under exclusive license to Springer Nature Switzerland AG 2023

D. Stockemer and J.-N. Bordeleau, *Quantitative Methods for the Social Sciences*, Springer Texts in Political Science and International Relations,
https://doi.org/10.1007/978-3-031-34583-8

(continued)

	MSP	ST	Gender	Year	TP	FWA	QECA	ATSP
Student 26	60	10	1	4	2	40	70	10
Student 27	50	11	0	3	2	30	50	0
Student 28	75	4	0	4	3	80	70	0
Student 29	80	7	0	6	4	90	30	0
Student 30	30	12	1	3	0	20	80	40
Student 31	70	7	0	4	3	70	30	10
Student 32	70	14	1	2	1	0	50	100
Student 33	70	3	0	4	3	60	40	50
Student 34	60	11	1	4	2	50	50	70
Student 35	70	9	0	2	1	60	20	40
Student 36	60	11	0	3	1	20	60	60
Student 37	60	11	1	4	1	30	40	20
Student 38	90	8	1	1	2	50	10	50
Student 39	70	9	0	3	3	50	90	30
Student 40	200	10	1	4	5	40	20	100

MSP = money spent partying; *ST* = study time; Gender = Gender; Year = Year; *TS* = times spent parting per week; *FWA* = fun without alcohol, *QECA* = quality of extracurricular activities; *ATSP* = amount of tuition the student pays

Appendix 2: Possible Group Assignments that Go with This Course

As an optional component, this book is built around a practical assignment. The assignment consists of a semester-long group project, which gives students the opportunity to practically apply their quantitative research skills. In more detail, at the beginning of the term, students are assigned to a study/ working group that consists of four individuals. Over the course of the semester, each group is expected to draft an original questionnaire, solicit 40 respondents of their survey (i.e., 10 per student), and perform a set of exercises with their data (i.e., some exercises on descriptive statistics, means testing/ correlation and regression analysis).

Assignment 1:

Go together in groups of 4 to 5 people and design your own questionnaire. It should include continuous, dummy, and categorical variables (after Chap. 4).

Assignment 2:

Each group member should collect 10 surveys based on a convenience sample. Because of time constraints there is no need to conduct a pre-test.

Assignment 3:

Conduct some descriptive statistics with some of your variables. Also construct a pie chart, boxplot and histogram.

Assignment 4:

Your assignment will consist of a number of data exercises.

1. Graph your dependent variable as a histogram
2. Graph your dependent variable and one continuous independent variable as a boxplot
3. Display some descriptive statistics
4. Conduct an independent samples t-test. Use the dependent variable of your study; as grouping variable use your dichotomous variable (or one of your dichotomous variables).

© The Editor(s) (if applicable) and The Author(s), under exclusive license to Springer Nature Switzerland AG 2023

D. Stockemer and J.-N. Bordeleau, *Quantitative Methods for the Social Sciences*, Springer Texts in Political Science and International Relations,
https://doi.org/10.1007/978-3-031-34583-8

5. Conduct a One-Way ANOVA Test. Use the Dependent Variable of Your Study. As a Factor, Use One of Your Ordinal Variables.
6. Run a correlation matrix with your dependent variable and two other continuous variables.
7. Run a multivariate regression analysis with all your independent variables and your dependent variable.

Book Summary

This textbook offers an essential introduction to survey research and quantitative methods. Building on the premise that statistical methods need to be learned in a practical fashion, the book guides students through the various steps of the survey research process and helps apply those steps toward a real-life example.

In detail, the textbook introduces readers to the four pillars of survey research and quantitative analysis in the social sciences: (1) what is survey research and why it is important, (2) preparing a survey, (3) conducting a survey, and (4) analyzing a survey. Students are shown how to create their own questionnaire based on theoretically driven hypotheses to achieve empirical findings from a solid survey dataset. Lastly, readers are introduced to data analysis to test their hypotheses using bivariate and multivariate statistical methods.

The book covers the theoretical, logical, and mathematical foundations of univariate, bivariate, and multivariate tests. In addition, it provides clear instructions on how to conduct these analyses using the R programming language. Given the breadth of its coverage, this textbook is suitable for introductory statistics, survey research, as well as quantitative methods classes in the social sciences.

© The Editor(s) (if applicable) and The Author(s), under exclusive license to Springer Nature Switzerland AG 2023

D. Stockemer and J.-N. Bordeleau, *Quantitative Methods for the Social Sciences*, Springer Texts in Political Science and International Relations,
https://doi.org/10.1007/978-3-031-34583-8

GPSR Compliance
The European Union's (EU) General Product Safety Regulation (GPSR) is a set of rules that requires consumer products to be safe and our obligations to ensure this.

If you have any concerns about our products, you can contact us on

ProductSafety@springernature.com

In case Publisher is established outside the EU, the EU authorized representative is:

Springer Nature Customer Service Center GmbH
Europaplatz 3
69115 Heidelberg, Germany

www.ingramcontent.com/pod-product-compliance
Lightning Source LLC
LaVergne TN
LVHW011758140925
821067LV00004B/5
* 9 7 8 3 0 3 1 3 4 5 8 2 1 *